IT GOES WING

IT GOES WITHOUT SAYING

TAKING THE GUESSWORK OUT OF
YOUR PHD IN ENGINEERING

CAROLINE BOUDOUX

FOREWORD BY ERIC MAZUR

THE MIT PRESS   CAMBRIDGE, MASSACHUSETTS   LONDON, ENGLAND

This book was set in StoneSerif and Avenir by Westchester Publishing Services. Printed and bound in the United States of America.

Library of Congress Cataloging-in-Publication Data

Names: Boudoux, Caroline, author. | Mazur, Eric, writer of foreword.
Title: It goes without saying : taking the guesswork out of your PhD in
   engineering / Caroline Boudoux; foreword by Eric Mazur.
Description: Cambridge, Massachusetts : The MIT Press, [2024] | Includes
   bibliographical references and index.
Identifiers: LCCN 2023037434 (print) | LCCN 2023037435 (ebook) |
   ISBN 9780262548205 (paperback) | ISBN 9780262378994 (epub) |
   ISBN 9780262378987 (pdf)
Subjects: LCSH: Engineering—Study and teaching (Graduate)—Handbooks,
   manuals, etc. | Engineering—Research—Management—Handbooks, manuals,
   etc. | Doctor of philosophy degree—Handbooks, manuals, etc. |
   Dissertations, Academic—Handbooks, manuals, etc. | Doctoral
   students—Vocational guidance.
Classification: LCC T65 .B78 2024 (print) | LCC T65 (ebook) |
   DDC 607.1/1—dc23/eng/20231205
LC record available at https://lccn.loc.gov/2023037434
LC ebook record available at https://lccn.loc.gov/2023037435

10 9 8 7 6 5 4 3 2 1

To Émile and Mario, it goes without saying.

"Be yourself, no matter what they say."
Sting

CONTENTS

LIST OF FIGURES

LIST OF TABLES

FOREWORD

Academics tend to suffer from what Susan Ambrose, in her book *How Learning Works*, calls "the expert blind spot"—once we acquire a new skill or knowledge, we tend to forget the struggles we faced before reaching that expertise. It becomes so obvious that it, well, *goes without saying*. While the pathway to a PhD may be clear to PhD advisors, it may feel like a voyage to the unknown for the beginning (or aspiring) PhD student. The seasoned academic knows that progress in any doctoral research goes hand in hand with (productive) failure. In contrast, the novice PhD student often sees failures as setbacks rather than stepping stones to success. In this gem of a book, written with wit and full of practical examples, Caroline Boudoux illuminates and demystifies the journey to the PhD. What are the expectations of a PhD? How will a PhD affect my career and earnings? How does one find an advisor? How does one even find joy in the process?

Boudoux systematically tackles these questions, providing all the advice rookie PhD students need in a useful and accessible format. She begins by discussing what a successful PhD entails, what milestones to expect, how to put together a thesis proposal, and finally, what career options lay ahead after the hooding ceremony. The central chapters dive into the nuts and bolts of defining, planning, organizing, and managing a research project, and, finally, writing and defending a thesis, "the official vessel to disseminate new knowledge" in Boudoux's words. The book's final part begins with a chapter on perhaps one of the most central skills required to obtain a PhD: writing. Writing and reading scientific papers is an art. Unless we explicitly help students develop the necessary skills, academic communication will remain a significant burden for both PhD students and their advisors. In five succinct sections, Boudoux provides tips on structuring and writing a scientific paper that will likely improve the lives of many PhD students and their advisors. She concludes her field guide with strategies to overcome common hurdles, including lack of representation, unconscious bias, impostor syndrome, mental health issues, psychological safety, and becoming a caregiver while working on a PhD.

The pursuit of a PhD is a remarkable journey, a journey that challenges intellectually, emotionally, and, at times, even physically, as the demands of a PhD can be all-consuming. Boudoux's book guides PhD students through the ups and downs of the journey, offering insights, advice, and strategies to help navigate the challenges that come with advancing beyond the frontier of knowledge. Boudoux frequently draws from her experience as a graduate student and as a mentor to many to complement this guide with personal anecdotes that make the advice engaging and concrete. This makes the book a valuable resource not only for PhD students but also for their advisors. While her book is aimed at the engineer, the advice and practical strategies provided are much more broadly applicable, and the book should really be on the reading list of any beginning PhD student. *It Goes without Saying* shouldn't go without reading!

Eric Mazur
Academic Dean for Applied Sciences and Engineering
Balkanski Professor of Physics and Applied Physics
Harvard University

PREFACE

Definition 1: Tintinologist *Someone who studies Les Aventures de Tintin, Hergé's fictional Belgian reporter traveling the world with his dog, Snowy, and his best friend, Captain Haddock.*

My parents were tintinologists. I grew up learning new words from Captain Haddock's colorful expletives. In *The Castafiore Emerald*'s original version, Tintin and Captain Haddock famously exchanged about some safety instructions:[1]

Captain Haddock   Boullu must have warned you, I suppose.
Tintin   No, he didn't, but it goes without saying.
Captain Haddock   I know! But it gets even better when you say it!

This dialogue applies to so many aspects of the doctoral journey. While completing a PhD, what may appear an obstacle to the candidate may seem trivial to the seasoned academic. Some would even say: "It goes without saying." However, in the moment's heat (and pressure), everyone wins when implicit notions are made explicit. Indeed, the right coaching often is the difference between a happy and a painful experience. In an ideal world, the thesis advisor would communicate all tips to their mentees. In reality, however, junior advisors practicing one-to-one mentoring often transition to senior lab director positions overseeing research operations akin to large tech companies' research and development (R&D) departments. With time, workload, responsibilities, and, perhaps, fading memory, some details slip through the cracks and do not reach the student on time, if at all.

As a thesis advisor myself, I plead guilty (if not the fifth). I believe I was somewhat of a decent mentor to my first student. Fifteen (plus) years into my academic carer, I know I have said all of it, and many times over, just not all to every newest member of my lab. With this book, I wish to remedy the situation. This is indeed the book I wish my students would read at the onset of their doctoral journey. This is also the book I wish I had as I began my own journey. I had the privilege of being my advisors' first student—everything they knew, they (tried to) pass it on to me. And while they did a fantastic job communicating these tips to me, I can't say that I grasped all of them in a timely manner. Some tips had to be repeated and often made sense only a few years down the journey. I also joined their lab with my own *a priori* conceptions about research, my own culture, which arguably was not so far from theirs (*et pourtant!*), and my own understanding of what a Doctor of Philosophy (PhD) is about, which, at times, was hard to reform.

The bits of wisdom in this book come from my own experience as a graduate student at the Massachusetts Institute of Technology (MIT), from current literature, from mentoring graduate students, and teaching mandatory doctoral workshops at Polytechnique Montréal (PolyMtl).[2] Individually, each bit is relatively trivial. As a whole, this book serves as a basis. Complete? Perhaps not. However, each student has resonated with some of its dimensions. I have taught with brilliant colleagues and, more importantly, to impressive junior scholars who have, through their questions and comments, challenged my views of the nature of a PhD in science, technology, engineering, and mathematics (STEM). Here, I submit examples taken from my journey and theirs. I am not advocating for a unique, cookie-cutter path. I merely suggest a common vocabulary to foster discussions between students and their advisors and to empower candidates to ask questions, lead their research projects, and take control of their trajectories. You may agree with me or not: what matters is that important topics such as expectations, thesis proposal, predatory publishing, intellectual property, and mental health, to name a few, are discussed early, openly, and often.

I hope some aspects of this book resonate with you too and that it helps you lead a happy PhD.

1

MOTIVATIONS

"The first principle is that you must not fool yourself and you are the easiest person to fool."

Richard P. Feynman[1]

This book explores the important aspects of doctoral studies in engineering (and science in general) in the hope of confronting your expectations with the reality of your journey. Before reading this chapter, take a moment to consider the following question.

Box 1.1
Motivations

What are your primary motivations for completing a PhD? Write your answers down, and perhaps discuss your answers with a friend. Committing to an answer is a great way to fully engage in this conversation.[2]

Figure 1.1 shows answers from doctoral students in engineering and other STEM disciplines. Indeed, a doctorate allows you to gain the necessary skills to tackle complex problems, navigate previously unexplored territories, and become an independent researcher, either in academia, a government laboratory, or the private sector. Identifying a strong

1.1 Typical answers from PhD students to the question: What is your primary motivation for pursuing a doctorate in engineering?

motivation for pursuing your PhD—strong enough to get you through the years between the initial honeymoon period and the anticipation of finishing—is a factor for success in your studies.[3] Other not-so-convincing motivations did not make it to figure 1.1. Indeed, to paraphrase the title of the 1966 Italian spaghetti Western film,[4] there are good, bad, and ugly reasons to pursue a PhD in engineering.

The Good   A PhD in engineering is right for you if you are curious and find value in intellectual challenges, unanswered questions, and contributing to pushing the frontier of knowledge using complex tools. In the right environment, you will develop personal and professional skills opening several career doors, both within and outside academia.

The Bad   After spending so much time in school, you fear the job market and figure that registering for a doctoral program allows you to stay in school a little while longer; you are terribly insecure and feel that a title will finally dissipate your impostor syndrome; or you are a victim of peer or family pressure to complete a doctorate. As Richard P. Feynman[5] said, "You have no responsibility to live up to what other people think you ought to accomplish. I have no responsibility to be like they expect me to be. It's their mistake, not my failing."

The Ugly   Satisfying a misplaced genius complex (i.e., "everyone is wrong but you"); fulfilling the ambitions of others the same way some play hockey to fulfill their parents' dream of making it the to the National

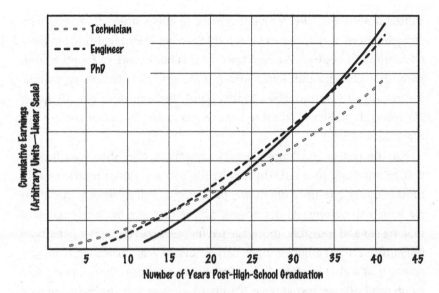

1.2 Lifetime earnings for a technician in engineering (high school plus three years), an engineer with a bachelor's degree (high school plus seven years), and a PhD in engineering (high school plus twelve years), assuming respective base salaries of X, $1.5X$, and $2X$, respectively, an annual raise of 3 percent and no ceiling on the annual salary. It also assumes that all degrees are completed free of debt.

Hockey League (NHL) [or National Football League (NFL), or *Fédération Internationale de Football* Association (FIFA), or International Cricket Coucil (ICC)], becoming rich and famous.

Incidentally, some of you might become rich, famous, or both, but making this your primary motivation is a proper path to disappointment. It is true that the average base salary for PhDs in engineering is slightly higher than the average base salary for engineers with a bachelor's degree, itself slightly higher than that of a technologist.[6] However, when factoring in the cost of the opportunity to spend several more years in school, the frugal stipend, and ever-increasing tuition, money should not be your primary motivation. Indeed, figure 1.2 simulates the lifetime earnings for a technician in engineering, an engineer with a bachelor's degree, and one with a PhD. The simulation assumes that all degrees are completed debt-free, somewhat of an optimistic assumption. Canadian considerations taint this simulation. In countries with exorbitant tuition and cost of living, the engineer and PhD curves of figure 1.2 must be shifted downward. The cumulative earning curve shows that pecuniary considerations should be second to intellectual ones regarding graduate education. Of course, you could use your school years to found a company that changes our world into a metaverse. While not impossible, statistically, it is improbable. To this point, the author of *The Algebra of Happiness*, Prof. Scott Galloway,[7] warns graduates, "Assume you are not Mark Zuckerberg."

For the rest of us, figure 1.2 says something else: there is a hidden cost to enrolling in a PhD. Indeed, while you are exploring the world's most exciting questions about *life, the Universe, and everything*: questions for which the answer is not always 42[8], your peers work as engineers, making car and, possibly, mortgage payments. An even higher penalty to pursuing a PhD in engineering occurs if, several years down the road, you realize that a PhD is, after all, not for you and opt out: time, money, and a potential title are lost at once. This book aims at fighting attrition: very many students indeed enroll in a PhD program never to graduate.

1.1 FIGHTING ATTRITION

Attrition can be fought uphill or downhill. If you are going to fail or leave, do so early. Indeed, figure 1.3 shows the distribution of doctoral degree durations in engineering, as reported by the Council of Graduate Education.[9] The histogram shows that one student out of twelve graduates within three years, one in ten graduate within their fourth year of registration, and that by the middle of their sixth year, more than half have completed their degree. The cumulative success rate curve plateaus at 65 percent, showing that, even ten years after entering a doctoral program, a third of students have not finished their PhD in engineering, and most likely never will.

Remark 1.1: Stipends   *Uneven funding practices also taint attrition numbers in engineering. While most doctoral candidates in engineering receive a stipend, there still exists a fraction of un- or self-funded students in science and engineering. While this fraction is typically smaller than in other disciplines, it contributes to the problem as research shows that funded students are more likely to graduate.*[10]

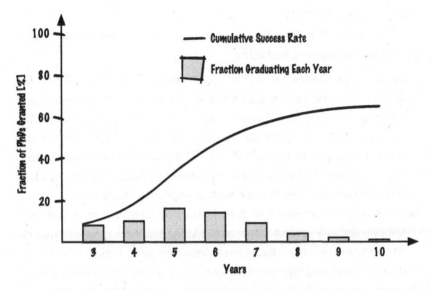

1.3 The duration of a PhD in engineering. The histogram shows the fraction of students graduating each year, with a cumulative curve plateauing at 65 percent, indicating a 35 percent attrition.

Finishing one's PhD is not the only path to changing the world. Stories of extremely successful engineering PhD dropouts do exist: Larry Page and Sergey Brin did leave their doctoral studies in engineering at Stanford University to found Google.[11] The purpose of this book is not to convince you either way. It is to equip you with the tools to make the right decision for you! In fact, "It is OK to quit your PhD" was the title of an article recently published in *Science*.[12] But, if you are undecided, make it a priority to make up your mind sooner rather than later. This book may help you understand what you are getting yourself into. The following chapters will uncover various project types, advising styles, and lab cultures to help tailor your doctoral experience to your expectations and ambitions.

Remark 1.2: Part-Time Studies   *Some institutions allow part-time studies, while others only allow students to register full-time. Many have a strict limit regarding the number of years allowed for completing a PhD degree. Avoid nasty surprises by finding out what your institution's rules are.*

CAUSES FOR ATTRITION

Funding aside, the hypotheses explaining student attrition revolve around several themes, including:

Grasp and expectations: A PhD in engineering may feel like life in *The Matrix*: "No one can be told what the Matrix is. You have to see it for yourself," said Morpheus to Neo in the 1999 movie.[13] Generally speaking, doctoral candidates poorly understand what a PhD is and what is expected of them.[14] Part I, "Doctoral Strategies in Engineering," explains what a doctorate in engineering is and exposes tips that often *go without saying* but are seldom explicitly mentioned.

Research project management: A metastudy by Geisinger et al. points at self-efficacy as a success factor in STEM graduate education.[15] Indeed, while PhD students are expected to lead a research project toward producing novel and significant results, they are seldom taught project management skills. Part II, "Leading a Research Project," teaches basic management skills within the context of engineering research.

Impostor syndrome and other incapacitating road bumps: Stress, exhaustion, and mental health issues,[16] as well as lack of writing skills,[17] are factors affecting attrition. Part III, "Tools of the Trade." explores a collection of modern practical themes ranging from writing tips to unconscious biases and mental health.

Student–Advisor Relationship: Across disciplines, the relationship between students and their advisors is cited as a critical factor for graduation.[18] Mismatch in expectations, lack of communication, and lack of support all affect belongingness. Strategies to improve this most important relationship are scattered throughout this book

1.2 TRAINING THE LEADERS OF TOMORROW

The Survey of Earned Doctorates from the US National Science Foundation (NSF)[19] shows an increase in the number of doctorates awarded by American institutions from 36,065 in 1990 to 55,283 in 2020. Of these, the fraction granted to engineering candidates increased from 13.6 percent in 1990 to 18.9 percent in 2020. In other words, in only thirty years, the number of PhDs in engineering doubled.[20] Yet, the number of PhD granting institutions only increased by 11 percent in the past two decades, suggesting that most PhDs do not become university faculty. Chapter 5 discusses possible career avenues and how to best prepare your professional portfolio while you are a student. Meanwhile, we faculty members must acknowledge that we are not training mini-me's anymore. Yes, it is our role to carry out the best research program to train PhD for research through research, but a well-rounded education must teach you more. Indeed, doctoral schools are training the leaders of tomorrow in many more spheres than just academia. We expect our PhD graduates to:[21]

1. probe their environment to target opportunities;
2. bring out original solutions having a significant impact;
3. mobilize intellectual, human, and material resources;
4. act and obtain results and transfer the benefits to society; and
5. exercise self-reflexivity to self-develop and to better contribute to progress.

This requires developing a family of useful skills revolving around:

content: scientific and technical knowledge and know-how;
process: methods for research and strategies toward innovation;
tools: communication and management; and
self: behavior and professionalism.

This books is based on class notes[22] developed for a series of mandatory doctoral workshops at PolyMtl aiming at developing:

- competencies not directly addressed through the research project;
- skills promoting access to a wider range of careers;
- factors increasing motivation; and
- ingredients improving the mentoring relationship between advisors and students.

In short, the aim of this book is to diminish attrition by clarifying expectations, providing you with better tools, notably in project management, and empowering you to make the next four or five (or even eight!) years among some of the best of your life.

1.3 HOW TO USE THIS BOOK

During my first week in Boston, I heard this old joke.[23] A student tries to pay for the full content of a grocery cart at a ten-item-and-under cash register. The cashier tells the student: "I cannot decide whether you are a humanities student who cannot count or an engineering student who cannot read." This book is for every student: it contains very few equations, and cartoons replace several thousands of words.[24] Yet, it was written with engineering and science students in mind: each chapter is short enough to be consumed during a subway ride taking you to the lab. Each chapter contains enough information to stimulate interesting conversations near the coffee machine when you arrive.

This book is structured in three parts, for a total of thirteen chapters:

Part I, "Doctoral Strategies in Engineering," defines what to expect from a PhD (chapter 2), describes a typical journey and common milestones (chapter 3), dissects sections of a thesis proposal (chapter 4), and prepares you for life after school (chapter 5).

Part II, 'Leading a Research Project," teaches notions of project management (chapter 6) through four steps called emergence and definition (chapter 7), planning and organizing (chapter 8), conducting and adapting (chapter 9), and concluding and submitting (chapter 10). Each chapter serves as an excuse to discuss important themes, such as sustainable development, time and priority management, ethics, risk management, and intellectual property management.

Part III, "Tools of the Trade," completes the toolbox with notions of scientific writing (chapter 11); a discussion on inclusion, diversity, equity, and accessibility; and the importance of maintaining one's mental health (chapter 12). Parenthood as a graduate student is also briefly discussed in (chapter 12).

It would be presumptuous of me to think that one will read this book from cover to cover in one sitting. The early chapters are most relevant to new graduate students (or students considering applying to a doctoral program), while others are more useful to seasoned doctoral candidates. However, each chapter is self-contained: peruse the table of contents, the index, or even the cartoons to find the topic that interests you the most, and come back to other topics as your PhD progresses.

AUDIENCE

This blueprint for a PhD was written by an engineer primarily for engineering students, yet much, if not all, of its content, would benefit other STEM doctoral candidates. These pages were also significantly tainted by my North American experience as a student and faculty member and by my French experience as a postdoctoral fellow. Despite PhDs being recognized worldwide, some variations exist in the process leading to the title. A generous network of international friends and colleagues has increased the reach of this work through remarks and comments. The endnotes highlight local flavors and should serve as a reminder to verify the rules specific to your institution.

I

DOCTORAL STRATEGIES IN ENGINEERING

2

A SUCCESSFUL PHD

"Nothing in life is to be feared, it is only to be understood."

Maria Skłodowska-Curie

Defining a successful PhD in engineering begins with understanding what a doctorate is in the broad sense before focusing on the specifics of engineering. Let us begin with a definition of the doctoral degree. The letters PhD stand for *philosophiae doctor*, Latin for doctor of philosophy, where *doctor* is Latin for teacher. It is universally recognized as a research qualification. A PhD is the highest academic degree a student can achieve.[1] PhDs are awarded across most academic disciplines, including engineering. In some schools, students may opt for an engineering doctorate (EngD or DEng) or a science doctorate (ScD or DSc), which is somewhat equivalent.

2.1 WHAT IS A PHD?

Box 2.1

Try Answering These Questions to Fully Engage in This Conversation:

- What are, in your opinion, the differences between a PhD and other university diplomas, such as master's or bachelor's degrees?

Box 2.1 (continued)

- One of the distinctive requirements for obtaining a PhD is to make an original contribution. What is an original contribution?
- What do you expect from your thesis advisor?
- What are, in your opinion, your advisors' expectations from their graduate students?
- How will you know that you are ready to defend your thesis?
- What are the job prospects for someone with a PhD?

The PhD differs from other doctorates [such as a medical degree (MD), or a doctorate of pharmacy (PharmD)] through the following characteristics:

- several years of independent research on a project;
- significant contributions to an original topic;
- mentorship from an academic expert called *the advisor*,[2]
- the preparation of a comprehensive *dissertation*; and
- a *thesis defense* against independent experts in the field.

In some countries, the PhD confers the title of Dr. However, where I live—the Canadian province of Québec—such a title is strictly reserved for MDs, veterinarians, and dentists. PhDs may only use their title after their names, as in: "Caroline Boudoux, PhD" or "Caroline Boudoux, doctor of nuclear engineering."[3]

Definition 2.1: Thesis   *The PhD thesis—also called a doctoral dissertation— is a document presenting the author's research. It is one of the requirements for completing a doctorate.*

The dissertation encompasses the thesis statement—the mental construct the candidate tries to prove or disprove—its associated context, and research-based arguments and conclusions. The thesis is evaluated by a jury—aka the thesis committee—ultimately deciding whether or not a student is deserving of a PhD.

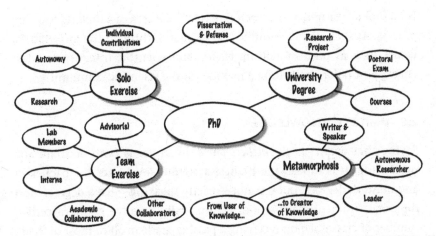

2.1 The many attributes of a PhD: a university degree, a solo exercise, a team exercise, and a metamorphosis. Almost never mentioned as an attribute of a PhD: it is a difficult enterprise.

ATTRIBUTES OF A PHD

Figure 2.1 maps the many attributes of a PhD degree as a:

university degree: the PhD candidate is responsible for navigating through a series of academic milestones and deadlines, including the successful completion of a comprehensive doctoral exam, research project, thesis defense, and dissertation. Chapter 3 explores the doctoral journey and typical milestones;

metamorphosis: from user to creator of knowledge, the PhD candidate is expected to acquire complementary skills such as autonomy, leadership, and communication. Section 2.3 explores the several facets of the metamorphosis;

solo project: the PhD candidate is expected to learn the ropes of research through realizing an individual research project leading to an original and significant contribution to the body of knowledge of their field, the defense of a thesis, and the completion of a dissertation. Section 2.4 discusses the concept of originality and significance; and

team effort: under the guidance of an expert, the PhD candidate is expected to contribute to research, often as part of a research team composed of other laboratory members, academic or industrial collaborators, thesis committee members, and actors of an international research community. Section 2.6 discusses respective expectations.

For a rookie researcher, the doctorate is the first step of a lifelong learning process. As such, the committee also evaluates, through the dissertation, values such as professionalism, ethics, and scientific integrity as a last checkpoint before becoming a member of the scientific community.

2.2 LIMITS OF KNOWLEDGE

Saying that a PhD is *the highest university degree* is both accurate and misleading. It is indeed the highest academic level after the bachelor's and master's degrees, but it is not—strictly speaking—just a degree. Most diplomas, from primary school to university, involve taking a certain number of classes. Upon receiving a passing grade in all or most of them, one reasonably expects to receive a diploma and, perhaps, a title. In most university degrees, passing all classes required by the curriculum guarantees the outcome. Not for a PhD: sitting idly through classes, handing in assignments, and writing exams is not enough. A PhD requires successfully defending an original thesis that will contribute to advancing humanity's quest for knowledge.

CIRCLE OF KNOWLEDGE

In his book *The Illustrated Guide to a PhD*,[4] Prof. Matt Might describes the notion of circles of knowledge. Figure 2.2 shows the role of each degree within the context of human knowledge. Imagine a circle containing all of humanity's knowledge. Primary school teaches pupils the necessary tools to learn more: reading and counting. Some general comprehension is also shared along the way, namely in sciences, geography, and history. The smallest circle represents knowledge acquired at this level. High school adds a layer of understanding, represented by a doughnut: a few skills in most fields—languages, mathematics, natural and social sciences—that is, in all directions of the knowledge circle. A bachelor's degree adds an extra layer of general knowledge with a specialty resulting in a bud sprouting radially in one direction. While a master's degree nourishes this specialty, a profound literature search during the early phase of a PhD allows for reaching the boundary of human knowledge. Where human knowledge ends, the PhD begins. Figure 2.3 zooms in at

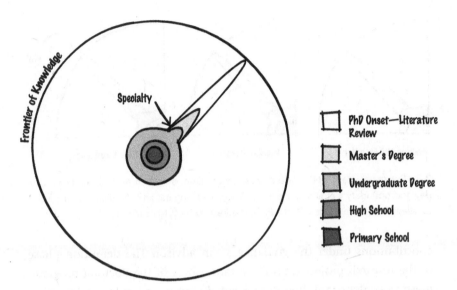

2.2 The circle containing all human knowledge. The role of primary and high schools is to provide general education. A bachelor's degree is still pretty general, but students also acquire a specialty—represented by a bud in a particular direction. The role of the master's degree is to specialize even more. In the first year of a PhD degree, students perform a literature review bringing them to the limits of what is known: the frontier of knowledge. Inspired by Prof. Matt Might.

the tip of knowledge, where it meets the unknown. Original hypotheses, adequate simulations, and perhaps careful experiments all contribute to one's effort to push against the frontier until, one day, the boundary yields and human knowledge expands. The dent created in the circle is a PhD. All the (mostly figurative) blood, sweat, and tears, all those years of intense study contained within a tiny bump. The limits of human knowledge are receding one thesis at a time. The accumulation of all PhDs in all fields contributes to expanding the sphere of knowledge in all directions. A PhD is therefore not an accumulation of years spent in school. Instead, it is the displacement of the limit of knowledge in a specific direction [as shown in figure 2.3(arrow)].

MASTER'S VS PHD
One key difference between the two graduate degrees is that master's students implement while PhD candidates make original and significant

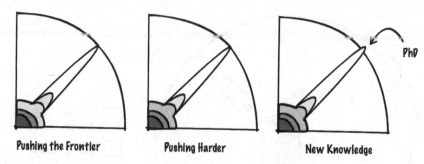

Pushing the Frontier   Pushing Harder   New Knowledge

2.3 For many years, PhD students push at the limit, and then push some more. One day, the boundary yields and a dent is made. This original and significant contribution to scientific knowledge is called a PhD. Inspired by Prof. Matt Might.

contributions under the guidance of an advisor. The definition phase of the research project is an important aspect of the doctoral program: master's students are typically provided with a project, as an initiation to research—a first glimpse of a complex world. Master's students may make important contributions to their research fields, but this is not an expectation nor a requirement for such a degree. Master's students produce and *present* a memoir; PhD students produce and *defend* a thesis. In addition, at the PhD level, there is an expectation of obtaining original and significant findings. In most cases, there is an expectation to publish such findings.

2.3 METAMORPHOSIS

Doctoral studies involve the student's metamorphosis from user to creator of knowledge. This transformation occurs on several fronts, as shown in table 2.1. Before the PhD, the acquaintance between students and science happens through didactically prepared material. A teacher or a *supervisor* carefully charters the student's path to enlightenment. During the PhD, the student creates new knowledge under the guidance of an *advisor*. Together, they create a new path through uncharted territories. For the first time, the student is being asked questions for which there exists no known solution and, often, no known strategy.[5]

Table 2.1 The metamorphosis of the graduate student[a]

Before a PhD		During and after a PhD
Using existing knowledge	→	Creating new knowledge
Applying known theories, accepted approaches, and validated tools	→	Venturing into unfamiliar terrain, identifying promising paths to new knowledge, and validating their results
Progressing within a strict frame and short timeline	→	Planning one's progress with no rigid timeline, despite academic or personal deadlines
Working under a supervisor	→	Working for, then with, and, eventually, surpassing an advisor

[a] Inspired from Jean Nicolas, "Réussir son doctorat" (Class Notes, 2012)

Box 2.2
Independence vs Lifelong Friendship

The goal is to achieve intellectual independence in order to eventually lead research endeavors. This is not incompatible with forming a lifelong research partnership with your advisor. Recent Nobel laureates Paul Milgrom and Robert Wilson are such a pair. Not only are the two laureates still working together but they are also neighbors, as witnessed by Milgrom's security camera the night they received a call from Stockholm.[6] In such a partnership, it is paramount that, in addition to the shared knowledge, each member of the pair develops his or her own expertise.

Table 2.2 illustrates the evolution of competencies of the graduate student from master's to PhD. At the PhD level, students are required to develop autonomy, one very important competency. An important point to stress is the gradual transformation of the research supervisor from advisor to future colleague.

AUTONOMY

PhD students in any field must work autonomously and expertly on a research project leading to original and significant contributions to knowledge or development. Competency elements associated with autonomy are:

Table 2.2 Evolution of the graduate student's competencies[a]

Competency	Master's	PhD
Conduct rigorous research …	under the supervision of an expert.	Autonomously and expertly lead the project.
Contribute to …	advancing knowledge under the close supervision of an expert.	an original and significant advancement to knowledge in collaboration with an adviser.
Identify, manage, and analyze information relevant to …	a research project.	a research field.
Communicate …	the results of the research project …	across a wide range of situations, findings from your project, along with their impact on the field.

Respect standards, rules of ethics and fairness, and best practices for research.

Commit to a process of lifelong learning and improvement.

[a] Inspired from Jean Dansereau and et al., *Competencies, Competency Elements, and Resources to Mobilize for Graduate Studies, Professional Master's, Research-Based Master's, and Doctorate*, https://www.polymtl.ca/renseignements-generaux/en/official-documents, 2014.

- justifying a research problem through an exhaustive literature review;
- formulating a research question, an original hypothesis, and specific research objectives;
- selecting, adapting, and designing research methods or analysis techniques;
- establishing (and following) a realistic timeline taking contingencies into account;
- analyzing, interpreting, summarizing, and evaluating results within the context of existing scientific literature; and
- evaluating the impact of the research project.

While the transition from master's to PhD student does occur overnight, one cannot expect instantaneous transformation into an autonomous and independent researcher. As an advisor, my expectation regarding independence and autonomy is illustrated in figure 2.4. Through my research partnership with PhD candidates, the ratio between a student's effort (S) and mine as a primary advisor (A) evolves as follows:

first year: $S < A$ The advisor proposes the first research question and objectives, secures the materials and designs the early method, assists with result analysis, and participates actively in drafting and revising a manuscript. The student catches up with literature, assembles the experimental setup, acquires data, and gets acquainted with the scientific writing style of the advisor;

middle years: $S \approx A$ The advisor and the student agree on the following research question and objectives. At this point, the student has had to preliminarily defend their ideas and plan in front of a panel of examiner in an event called the comprehensive doctoral exam. The student proposes a strategy for the implementation, acquires and analyzes data, and prepares a complete draft for the advisor; and

final year: $S > A$ The student proposes novel research avenues and objectives, a strategy for the implementation, acquires and analyzes data, and prepares a complete draft for the advisor. The advisor assists in securing the resources, perhaps involves the student in grant writing, and revises manuscripts.

Figure 2.4 also shows two other trajectories: that of the prodigy (for whom $S > A$, from the start) and the late bloomer. The autonomy of the prodigy is

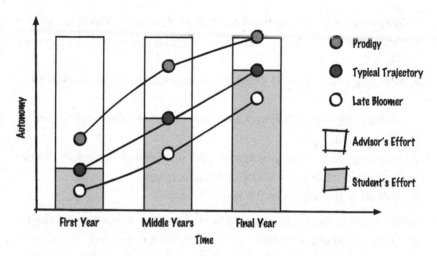

2.4 Shared effort between the advisor and the student as autonomy grows: the typical trajectory (middle), the late bloomer (bottom), and the prodigy (top). Not shown is the horizontal line of the candidate who flatlines, or fails at developing the required level of autonomy.

high to start with and further grows, sometimes exponentially: the dream student in the minds of many. The late bloomer's independence, on the other hand, begins low and improves, but not as rapidly as one would hope.

Box 2.3
Late Blooming

If you find yourself following the late bloomer's path, ask yourself: "Why is this?" Do you suffer from:

Lack of knowledge? Perhaps you could attend some workshop, take extra classes, online, at your school, in a summer school, find a collaborator with complementary expertise.

Lack of self-confidence? Is the problem too complex? Can you break it down into smaller, easier-to-manage parts?

Lack of motivation? Do you find yourself procrastinating a lot? Are you working on the right aspect of the project?

Box 2.3 (continued)

Lack of specific short- and long-term goals? Is the project not ambitious enough? To quote a Marvel character: "if you aim at nothing, you hit nothing."[7] Or perhaps your project is ill-defined and you cannot seem to organize proper activities around it (read part II for tips on project management).

Whatever the reason for blooming later than you would like, try to identify and address the root cause. Make a plan and discuss it with other students around you until you feel confident enough to bring it up with your advisor. Hearing about your challenges is part of the advisor's job description. However, contributing to a personalized solution is more straightforward when suggestions for improvement come from you. Be proactive, and, as recommended in figure 2.5, help us help you.

2.4 CONTRIBUTIONS

The output of a PhD consists in *original and significant contributions* to the sphere of knowledge. Ling and Yang[8] represent the path to the original contribution as a series of curves on a 2D graph, shown in figure 2.6. In such a graph, the x- and y-axes represent, respectively, the breadth and depth of expertise within a certain field. The horizontal line represents the frontier of knowledge. As in figure 2.2, coursework and literature review allow you to approach the state of the art of research. In the first year of your PhD, you explore some topics and advance your understanding of these exploratory topics. Soon, you are required to define a thesis proposal based on your most promising preliminary research, and this proposal becomes your primary, if not sole, topic. With your contribution, this topic is researched to the point where you push the frontier of knowledge a little further. Each curve roughly corresponds to a year, and the area under the curve represents your effort. Figure 2.6 may also help to explain what a PhD is not:

2.5 Help us (faculty) help you (the students). Be proactive in searching for solutions to the problems you face. You coming up with ideas shows your dedication and might even inspire new ones.

- a Dirac delta function showing a direct path to the original and significant contribution with no area under the curve (AUC), that is, no effort; or
- a series of peaks never reaching the frontier of knowledge: a PhD is not a series of master's-level projects.

ON QUANTITY

As engineers, we love numbers. It is thus no surprise that many have seen one form or another of the equation

$$1 \text{ doctorate} \overset{?}{=} 3 \text{ papers.} \tag{2.1}$$

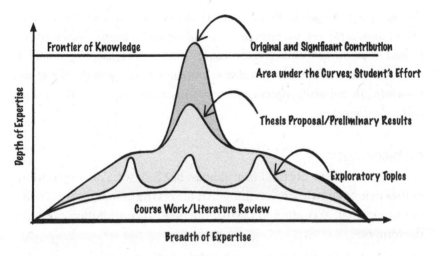

2.6 From topics to thesis. Focusing on a very specialized subject leads the PhD candidate to become an expert in a particular field.

Indeed, many of us, faculty, are guilty of describing the doctoral journey as *a specific number of contributions* to the field. I want to pause for a moment and stress that I am yet to find in the (yet-to-be-written) *Big Book of Doctoral Rules* a line stating that a doctoral degree equal three publications, conference presentations, patents, or combinations of these. Some may graduate with no publications aside from their dissertation, others with a single published article, and some with an astounding number of manuscripts. Even within engineering, every specialty treats publications differently. In some fields, contributing an oral presentation to an international conference is a walk in the park, while for others, it is akin to a Spartan race. In some journals, the publication time is as short as a few weeks, while in others, the reviewing process alone can last years. Overall, the doctoral contribution should be judged in and of itself by a panel of experts and not based on a single bibliographic metric. That having been said, I find it helpful to sketch the path of a new student as being centered on three themes, with progressive levels of risk and autonomy, as previously described in figure 2.4. The research road map focuses on knowledge creation rather than the number of containers (i.e., publications) of knowledge.

Definition 2.2: Negative Proof *Proving that something does not exist is almost impossible, yet, we find ourselves making such statements quite often. Instead of saying that no university requires three publications to graduate, I should have used a common formulation in scientific writing: to the best of my knowledge, no university requires having published three papers as part of their doctoral requirements.*

ON ORIGINALITY

How original must a doctoral contribution be? Following our description of the circle of knowledge, anything that moves the limit outward is considered original. According to Lovitts,[9] an original contribution can take the form of:

- a piece of work that has never been done before;
- a concept that can be patented;
- an effort that is inspired and creative;
- a technique that paves the way for others, opens new horizons;
- a project that is publishable;
- a solution to a previously unsolved problem;
- a new idea that breaks paradigms; or
- a theory that contradicts established thinking.

It is important to recognize that sources of originality may be found in every step of the scientific—or engineering—method, as we will discuss in chapter 7.

ON SIGNIFICANCE AND IMPACT

For many, doctoral studies equate putting life on hold for a few years, with the implicit consequence that the results be worthy of the sacrifice. How does the community judge significance and impact?

Box 2.4

An Original and Significant Contribution

In your field of research, what are the attributes of an original and significant contribution? Discuss your answers with PhD students in your cohort or in your lab. Consult previous PhD theses from alumni from your research group.

In addition to being original and novel, the content of the thesis results should affect how the community solves a problem, how people see things, or how people think: "The key feature of a significant contribution is that it has consequences" (Barbara Lovitts, 2007[10]).

In engineering, significance and impact develop around several axes. A significant contribution includes:[11]

- a new theory or a new model perhaps based on new observations;
- results that are publishable and, ideally, cited once published;
- findings that are important to many people; and
- an application that affects people's lives and has commercial and/or industrial impact.

Significance is somewhat proportional to the size of the community affected by the results. At the lowest level, a doctoral thesis only affects members of the laboratory that hosted the research. At the other extreme, highly significant doctoral dissertations offer complete breakthroughs and create new paradigms.

Box 2.5

Extreme Impact

Paul Dirac's (hand-written!) PhD dissertation titled "Quantum Mechanics"[12] at the University of Cambridge in 1926 contributed to the development of two new fields in physics: quantum mechanics and quantum electrodynamics. Not only is the consequence of Dirac's doctoral work still felt almost one hundred years later, but in 1933, he was awarded the 1933 Nobel Prize in Physics, along with Erwin Schrödinger. One may say that the breadth of his work was anything but a Dirac delta function!

Most theses do not lead to immediate and fundamental paradigm shifts in the field. A small but sound contribution to a well-circumscribed topic makes for a great thesis. Furthermore, the timing of research affects its impact: it is much easier to measure the immediate impact on a hot topic than in some theoretical fields.

> **Box 2.6**
>
> **Maria Goeppert Mayer's Prediction**
>
> In her 1930 doctoral thesis,[13] Maria Goeppert Mayer predicted two-photon absorption by molecules, a theory that was only verified in the 1960s[14] with the first sufficiently powerful light sources (i.e., lasers), and fully exploited commercially another three decades later in two-photon laser scanning fluorescence microscopy.[15] As the 1963 Nobel Laureate in Physics, she is the second women in only five[16] to have been presented with the prestigious award.

Another, perhaps more ubiquitous, example of delayed impact is artificial intelligence (AI), predicted in 1950 by Alan Turing,[17] yet only deployed after decades of investments toward reducing computing costs.

STRATEGIES FOR IMPROVING SIGNIFICANCE AND IMPACT

The first obvious strategy to improve the significance of your work and fight insularity is to publish your work.[18] Before publishing in a scientific journal, parts of your work may be presented during your group meetings, at conferences (local and international), or at local workshops. Discussing your work in as many circles as possible allows for feedback from peers, teaches you adapt your communication approaches based on your audience and helps mitigate the risk of a bad surprise from reviewers or thesis committee members. As an added bonus, you are constructing your network of peers, which is necessary for both feedback and career development.

Remark 2.1: Confidentiality Circle *Always consult your advisor before disclosing part of your work outside your immediate research circle to ensure that the intellectual property (IP) protection strategy is respected. Presenting your work during group meetings allows for honing one's communication styles while discussing ideas in a friendly environment, typically with no IP considerations.*

ON METRICS

How do scientists measure significance and impact? Because we are numbers people (perhaps a euphemism for geeks), we have a tendency to measure significance and impact with plenty of metrics. In this section, we highlight the pitfall associated with reducing significance and impact to a single number, such as, for example:

dollar amount: by tallying how much money an idea generated in research contracts and grants? A drawback of this method is that it sometimes favors very applied research, to the detriment of fundamental research;

number of scientific publications: such an approach may lead to splitting research topics into small letter-type contributions to the detriment of larger publications in more influential journals. The technique is pejoratively referred to as salami slicing or salami publishing;[19] and

bibliometric indices: bibliometry refers to the use of mathematical and statistical methods to study and identify patterns in the usage of materials and services within a library or to analyze the historical development of a specific body of literature, especially its authorship, publication, and use.[20] In addition to the total number of citations, two metrics, in particular, are widely known and worth discussing: the impact factor (IF) and the h-index, aiming to quantify the respective influences of journals and researchers.

Definition 2.3: Impact Factor *The IF of a journal is a dynamic index that records the number of citations in a given year (say the year 2020) to its articles published during the previous two years (e.g., articles published in 2018 and 2019) divided by the total number of articles published by the journal during the same previous two years. Citations may come from articles from all journals, including the one for which the IF is being calculated.*[21]

As is it the case for any index, the IF is criticized. For example, a very influential publication (being referenced, say, more than one hundred times) disproportionately boosts the arithmetic mean that is the IF.[22] Additionally, citation practices vary from one field to the next, resulting in a broad range of IFs depending on whether you are a mathematician or a biomedical engineer, for example.[23] Other indices, each with their own pros and cons, are also used to evaluate the impact of a journal. Their inventory is, however, outside the scope of this work. When trying to evaluate the past influence of a journal, keep in mind the following:

- An article published in a journal having an IF smaller than one has had, on average, less than one citation within the first two years of publication. Your paper, of course, may by the one positively affecting that journal's IF, or you may elect to give your work a better chance at being discovered by others.
- The past IF of a journal is never a guarantee that your paper will be cited: you are still required to produce quality science in order to get cited.
- The scope and mandate of a journal are often more important than its IF or any other metric.

When judging the quality of a research article, no index beats reading the paper and forming your own opinion, even when the IF of that journal is low.

Box 2.7
San Francisco Declaration on Research Assessment

In 2012, at the Annual Meeting of The American Society for Cell Biology, a group of scholars drafted the Declaration on Research Assessment (DORA),[24] which has since been signed by more than twenty-four thousand scholars from 164 countries. The declaration urges academia to eliminate journal-based metrics in the evaluation of funding proposals, appointments, and promotions, and to evaluate research on its own merit. Despite being more than ten years old, DORA has not percolated through all layers of evaluation committees just yet, and, as a result, many researchers still nurture their indices.

THE COST OF PUBLISHING RESEARCH

Research results are validated through a process called peer review. When a manuscript is submitted, the editor first verifies its alignment with the journal's topics. When the work is qualified as suitable by the editor, several researchers working on similar topics are invited to review the manuscript. These peers volunteer their time as part of their service to the scientific community and in recognition that when they submit manuscripts others will do the same.

> **Box 2.8**
> **Peer Review**
> How many publications should a researcher evaluate each year? One should review at least the yearly number of scientific publications published in one's lab divided by the acceptance rate of the journal in which one publishes.

Given that papers are reviewed for free, one may ask what is the cost of publishing? In addition to the salaries of the editing staff, journals must pay for archiving their publications, which includes maintaining a park of secure electronic servers.

A recent trend in scientific publishing is shifting the cost of publishing from the reader to the writer. Previously, university libraries would pay significant subscription costs to access articles otherwise locked behind a firewall. Others, not affiliated with large research institutions, had access to scientific knowledge through a costly pay-per-view system. Now, most funding agencies insist that the work they support be published in open-access journals or made available in public repositories within a set time frame.

Definition 2.4: Open Access   *"Open access literature is digital, online, free of charge, and free of most copyright and licensing restrictions."*[25] *This important initiative allows free knowledge transfer from researchers to humanity. It also shifts the burden of paying for publications from the reader to the authors of scientific work.*

While this new paradigm should be praised for having dramatically democratized research results, it also opened the door to predatory publishing.

Definition 2.5: Predatory Journals   *Predatory publishing is a deceptive publishing business of charging publication fees without providing adequate peer revision and credibility to a manuscript, thus producing science of questionable quality. Within five years of publication, 60 percent of articles published in predatory journals received no citations, as opposed to only 9 percent in legitimate ones.*[26]

Predatory publishing thrives from the notion that young researchers must publish if they don't want to perish. Such journals will aggressively solicit contributions, often with short delays and promises of immediate publishing. They also feed from the naiveté of new (and not so new) researchers—forewarned is forearmed!

Box 2.9
Predatory Publishing

How do you know whether a journal is legitimate or not? Here is a non-exhaustive list of clues:

- Legitimate journals are published by renowned publishing houses.[27]
- Equally legitimate journals are published by international or national technical societies.[28]
- Known predatory journals are compiled on blacklists found online.[29]
- Predatory journals will aggressively solicit contributions, often with short delays and promises of immediate publishing.

There is little doubt that, before long, you will be solicited to contribute a review paper for some journal. Before you do, consult your advisor (and certainly your librarian) to verify the journal's legitimacy. Other resources include the initiative Think.Check.Submit,[30] providing a checklist for rookie (and possibly also mature) scientists. Researchers may also become the subject of bibliometric ranking through the h- and i10-indices.

Definition 2.6: h- and i10-indices *For a researcher, the h-index—developed and named after Jorge E. Hirsch[31]—indicates the maximum number, h, of publications having been cited at least h times. Its cousin, the i10-index, compiles the number of publications having been cited at least ten times. Interestingly, Hirsch subsequently discussed the h-index's severe unintended consequences.[32]*

Neither the h- nor the i10-indices distinguish between self-citations and citations from other groups,[33] and both are biased toward longevity (i.e., senior versus junior scientists), as it often takes several years for a paper to inspire the work of others and for others to publish and cite their source of inspiration. The body of literature critiquing the use of such indices is ever growing;[34] nonetheless, graduate students should be aware of such rankings as they are often and unfortunately so used to rank scholarship applicants and faculty candidates. Hirsch argues that the h-index offers some predictive power.[35] However, keep in mind that Einstein, having published only 105 articles, is limited to an h-index of 105, no matter how influential his work has been (and still is).

Another limitation of counting the number of citations to evaluate impact concerns work that is so influential that it becomes general knowledge before accumulating citations. No one cites Isaac Newton's papers anymore.[36] In box 2.4, we discussed a seminal paper by Maria Goeppert Mayer, whose impact was only truly felt sixty years after publication. Her original paper from 1931 accumulated a few hundred citations, while that from subsequent work in 1991 amassed over ten thousand citations. Can you imagine a physics department chair saying: "We are sorry, Dr Goeppert Mayer, we cannot offer you a position because your number of citations is too low?"[37]

2.5 PROTAGONISTS

Figure 2.7 shows the typical stakeholders of a doctoral degree. In engineering, you will typically perform the research aspect of your degree in a laboratory, be it in a university, research center, or industry. Each environment comes with its own set of values and operating strategies. In a manner similar to corporate culture, these define the lab culture.

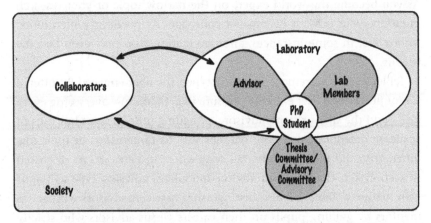

2.7 The main protagonists involved in a doctoral project. Distal players include members of the research community, who indirectly influence the course of the project through their publications.

Definition 2.7: Laboratory Culture   *Lab culture refers to the beliefs and behaviors that determine how laboratory members interact with senior members (post-doctoral fellows, technicians, and research assistants) and the laboratory director (aka the advisor) and handle research collaborations and production. As for corporate culture, the laboratory culture is often implied and develops organically over time from the cumulative traits of current and past laboratory members.[38] The lab culture will be reflected in expectations concerning working hours, contributions to research, level of mentoring, funding, travel support to conferences, requirements for teaching assistantship (TA) and research assistantship (RA), and so on.*

Laboratory cultures come in many flavors and are heavily tainted by the lab director. Your success and happiness throughout your studies are heavily influenced by the fit between you and the lab culture.

Definition 2.8: Advisor   *Someone who gives advice.[39] In the context of a PhD, an advisor stands somewhere between a supervisor—whose role is to oversee the work of a PhD candidate—and a mentor who, in addition to providing advice, shares expertise on a large variety of topics related to one's work and career.*

Your interaction with your thesis advisor falls somewhere within the boss–colleague spectrum. Initially, your thesis advisor closely supervises your work. As you gain autonomy, your thesis advisor becomes a mentor. As you become the world expert on the narrow topic of your research question, your advisor becomes a colleague. As previously mentioned in box 2.3, in some instances, after graduation, you may even become friends.

Table 2.3 lists extreme supervising types: the über-engaged PI (often a junior professor under extreme pressure to get tenure)—overseeing every task—and the sink-or-swim advisor—providing little support but tons of academic freedom. Each type handles various dimensions of their role differently, and most advisors fall somewhere in between. As discussed in section 2.3, the seasoned advisor modulates advising type as pupils gain maturity. Table 2.3 also distinguishes between advisors who act as mentors by guiding pupils on their careers versus advisors who strictly supervise research work.

Table 2.3 Extreme advising types[a]

Dimension	Über-engaged PI	Sink-or-swim advisor[b]
Ideas	mostly generated by the advisor	mostly generated by the student
Project	mostly defined by the advisor from a research grant or contract	mostly defined by the student through literature search
Supervision	regular standing meetings	sporadic meetings or on an as-needed basis
Failures	communicate as they happen	communicate when at least one fix is found
Publications	mostly written or largely edited and rewritten by the advisor	mostly written by the student
	Mentors	**Supervisors**
Metaphysics	are open to discussing "life, the universe, and everything"[c]	prefer a strictly professional rapport

[a] Nicolas, "Réussir son doctorat." (class notes, 2012).
[b] Roel Snieder and Ken Larner, *The Art of Being a Scientist: A Guide for Graduate Students And Their Mentors* (Cambridge University Press, 2009).
[c] Douglas Adams, *The Hitchhiker's Guide to the Galaxy* (Pan Books, 1979).

A survey of more than eight hundred PhD candidates in the Netherlands[40] showed that the mentor–mentee relationship, the candidate's sense of belonging, and the closeness between the dissertation topic and the advisor's core research program were positively related to satisfaction and negatively related with intentions to quit.

Box 2.10
What Is the Lab Culture?

If possible, before committing to a lab, ask to meet other graduate students to discuss where, between a micromanager and a sink-or-swim boss, does the lab director stand? How does this fit with you, honestly?

Remark 2.2: Adding a Co-Advisor *In some schools, students commit to a research lab on admission. When the lab culture fit is not optimal, consider adding a co-advisor to your advisory committee and choose someone*

with complementary skills and mentoring style. It is indeed common for stu-
dents, especially when involved in multiply disciplinary projects, to have a
co-advisor.

LAB MEMBERS

A research laboratory is a rich and, ideally, heterogeneous ecosystem. One
key in interacting with all lab members is to understand their respective
endgames. While there is no better way to know one's motivation than
to ask directly, here are some general directions:

undergraduate intern: seeks exposure to research, a great reference letter
for future positions, and possibly a coauthor spot on a manuscript;

master's student: is interested in contributing to furthering knowledge
without the long commitment associated with a PhD and is perhaps
still deciding whether or not to do enroll for the long run;

other PhD students: must too contribute to the advancement of the field
by navigating similar hurdles;

postdoctoral fellow: has a minimal time to establish a leadership position
on a narrow topic while learning every aspect of the trade not explicitly
taught in graduate school, such as some managerial skills; and

technician and research associate: as permanent members of the team,
their role is to train junior members, maintain the memory of the lab
and preserve a healthy working atmosphere.

Box 2.11
What Is Your Endgame?

Discuss this with your fellow lab members: What are their goals and ambitions? Share yours with them.

COMMITTEE MEMBERS

Graduate students are guided through their doctoral journey (see chapter 3) by several committees, including:

the thesis proposal committee: a jury comprising faculty members deciding on the merit of the thesis proposal;

the advisory committee: an optional committee of experts providing feedback on a semester or yearly basis; and

the thesis committee: the jury assembled to provide a verdict on the acceptability of the doctoral dissertation.

Remark 2.3: Thesis Committee *In some schools, a thesis committee is assembled at the onset of the thesis work and remains constant throughout the years. In other schools, a preliminary jury is assembled for the thesis proposal, and, depending on the evolution of the research project, a final, and perhaps different, committee is assembled to judge the dissertation and thesis defense. An advisory committee may thus be formed between the two events to help the student navigate the middle years of a PhD. An advisory committee is particularly useful when research involves partnerships with non-academic stakeholders to provide the students with a unbiased opinions.*

The various committees guide the student through the meanders of doctoral years from the planning phase—resulting in a thesis proposal—to the preparation and defense of the thesis. Their role and composition vary according to rules and guidelines specific to the student's institution. These committees can advise you on more than one dimension:[41]

- course work;
- comprehensive or qualifying exam;
- thesis proposal;
- research progress; and
- dissertation and thesis defense.

The candidate's research advisors are sometimes included in the thesis proposal jury and almost always part of the thesis defense jury. Other members are researchers (academic or industrial) who are academically independent of the research advisors.

Definition 2.9: Academic Independence *The definition of academic independence varies with institutions, but is typically defined as not holding joint funding or not having coauthored publications within the past five years.*

Remark 2.4: Approval *The composition of your thesis committees follows the guidelines of your institution and must be approved prior to scheduling oral exams, presentations, and the thesis defense. There is more on this in chapter 10.*

Jorge Cham[42], creator of the tragico-comic book series called *PhD Comics*, calls the thesis committee: "an impossibly difficult group to get together in one room but who nevertheless hold your future in their hands depending on their ability to reach a civilized consensus."

He is not entirely wrong. As independent experts, their points of view on the best course of action, their interpretation of your data, or their expectation of what a PhD is may differ. On the plus side, you gain a breadth of points of view. On the minus side, you need to satisfy the requirements of many before graduating. Do remember, however, that peer review is a constant theme of life in academia and, arguably, life as a researcher in general. Sarcasm aside, you should embrace the diversity of opinions provided by such a panel of experts. Try to filter out the noise of tone, cynicism, and sometimes, yes, arrogance to reach within the heart of the comment for constructive insight. When forming your committee, discuss with your advisor ways to promote a diversity of opinions. Avenues to explore include:[43]

expertise: a complex research project will benefit from the combined expertise of many experts;

relevance: industrial researchers, MDs or practitioners, in general, are great committee members ensuring the relevance of your engineering project to society, though they may not necessarily be well-versed in the intricacies of research or acquainted with the school's procedures;

network: members external to your department or to your school enhance your professional network; and

personal rapport: when choosing between two members, opt for the expert with whom you have a natural affinity to foster open and enthusiastic exchanges of ideas.

Remark 2.5: Inner Circle *Your thesis committee forms the core of your budding academic network. They are the chosen few who—in addition to your parents—will have read your dissertation.*

COLLABORATORS

Research is seldom performed by the mentor-mentee pair alone: several stakeholders are typically involved:

academic: researchers from other groups within a research center or between different institutions;

industrial: researchers from private companies through a collaborative endeavor or a research contract; and

clinical: practitioners (i.e., MDs and medical staff) from hospitals and clinical research centers.

Box 2.12
Stakeholders

Compile a list of all protagonists associated with your project. What are their motives for joining the team? What do they hope from this collaboration?

UNIVERSITY STAFF

For students, an academic department is a transient home. For university staff, it is their permanent workplace. It is your role to keep their work environment pleasant, despite all the pressure associated with (your!) deadlines, the photocopier eating your precious and original document, the internet bandwidth not being strong enough to support those important virtual meetings, and so on. Be known as the courteous person who sees people behind the cubicle. Perhaps, despite it being your responsibility to remember all deadlines and procedures, you might be issued a kind reminder and granted extra help for that form requiring signatures from five levels of the academic hierarchy. Recognizing that everyone contributes to the mission of the institution is the first step to fostering respectful and productive long-term relationships.

In chapter 9, we discuss the details of the kickoff meeting with all protagonists. In the meantime, we introduce common expectations.

2.6 EXPECTATIONS

"When students know the performance that is expected of them and the standards they will be judged on, they become more engaged in their intellectual and skills development, are better able to self-assess and correct deficiencies, and are better able to demonstrate what they know and can do."

Barbara E. Lovitts, 2007[44]

Expectations in graduate school are plenty. Society expects engineers to find new ways to turn natural resources into goods and services. Your parents might, at first, expect a Nobel Prize but eventually will settle for watching you walk in the graduation ceremony. Here, we focus on expectations from your school, your advisor, collaborators, and yourself.

FROM THE UNIVERSITY

In chapter 3, we discuss the requirements from the university as a series of milestones in the doctoral journey. While each university has different requirements, a common set of expectations involve that you:

- attend a certain number of courses and workshops;
- provide a study plan and defend a thesis proposal;
- pass a comprehensive exam;
- write and submit a dissertation; and
- defend your thesis.

Your university also expects you to follow a code of conduct, including guidelines on plagiarism and ethics. Make sure you familiarize yourself with such guidelines to avoid finding yourself in situations that leave indelible stains on your dossier.

Remark 2.6: Fine Prints *If you are funded through a scholarship, also familiarize yourself and comply with the expectations from the granting agency.*

FROM YOUR ADVISOR

In addition to academic considerations, your advisor may expect that you:

- help with teaching, writing articles, and applying for patents;
- apply to scholarships, travel grants, and other funding opportunities;
- build a prototype, an instrument, a sensor;
- develop an algorithm and write some code;
- create a database;
- invent and validate a new method;
- collaborate with fellow lab members; or
- coach more junior students.

Your advisor may also have specific expectations concerning meetings (regular weekly standing meetings versus by appointment), application to conferences, manuscript redaction, and coursework.

Box 2.13
Expectations

Using the list mentioned above as a starting point, discuss with your advisor your respective expectations. Further, verify with other laboratory members what is expected of them when applying for finding or submitting abstracts to conferences? How polished should a first draft be before discussing it? What is the policy regarding purchases, vacations, collaboration, authorship, and so on? You may also use the word cloud of figure 2.8 to verify which are the key attributes promoted by your advisor.

In an ideal world, lab policies should be explicitly written—the reality of a junior faculty starting up a lab often prohibits the creation of such documents. A good lab citizen (perhaps you?) could offer to help in that department.

Definition 2.10: Lab Book   *Students are expected to keep a lab book (also called a logbook) up to date. Ask your advisor what book to acquire and how to keep it up to date. Typically, the lab book is the property of the lab. Use the lab book not only to record experimental data but to chart ideas, to record a stream of consciousness. When things get tough, and you feel the well of ideas drying up, you can flip back to earlier times to measure your progress and follow up on roads not taken.*

2.8 Common expectations from advisors. Which attributes are prized by your lab?

Part of developing a good relationship with your advisor is sympathizing with their situation. You may see your advisor as a demigod on top of the world, academically speaking, but many are, in fact, akin to a duck paddling furiously to stay afloat. This is especially true of younger faculty before they are granted tenure.

Definition 2.11: Academic Tenure *A tenured position is an indefinite academic appointment that can only be terminated for a very limited set of reasons, such as for cause,[45] as opposed to a temporary position. Positions leading to tenure are called "tenure-track" positions. Chronologically, tenure-track faculty begin their careers as assistant professors before being promoted to associate— which roughly coincides with tenure, although some schools discuss tenure independently from ranking—and eventually to full professor, as illustrated in figure 2.9.*

Faculty submit their dossier for promotion to a committee comprising local and international academics. They are evaluated on the quality of their teaching (classes) and training (you), their research output (impact of their publications, number of successful grants, licensing of their

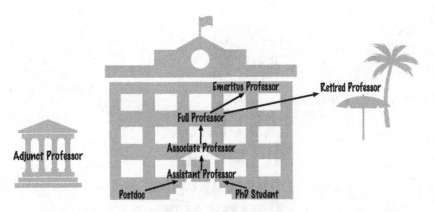

2.9 The typical academic journey from PhD student or postdoctoral fellow to recruitment as an assistant professor. The next promotion leads to being an associate professor, tenured or not, before finally being named full professor. Scientists from other institutions may apply for affiliation with a school and become adjunct professors. Some faculty retire from their role as a professor. A professor emeritus is a professor who has made significant contributions to their school while remaining active in research after retirement. The figure does not show the many (and most common) trajectories outside academia, which are discussed in chapter 5.

patents, and so on), and their service to their community (reviewer of theses, manuscripts, and grants, conference organizers, school committees, and so on). The time frame for assembling a dossier is relatively short given the diverse accomplishments expected of them, which explains why younger faculty might relay some of the pressure they feel onto their students. My goal in sharing this is to help create a bridge of understanding between students and their advisors. I was my advisor's first PhD student and recognize the perks that come with being the first student: almost unlimited attention, overflowing energy, and infectious enthusiasm for a state-of-the-art research project. Some prefer the quasi-anonymity of being part of a very large group coupled with the reputation of an established researcher. There is no ideal solution: what counts is building a good mentor-mentee working relationship.

FROM THE STUDENT
While all your school's and advisor's expectations are worthwhile, they may interfere with yours, which may include:

2.10 Common expectations from students, compiled from students attending doctoral workshops at PolyMtl. Which ones are the most important for you?

- complete the study program within a given number of years;
- support for prize and award applications;
- reach a work-life balance;
- build a professional network;
- develop a strong scientific and technical portfolio;
- completing an industrial internship;
- acquire professional skills; and
- become a proficient technical writer.

Other common expectations are listed in figure 2.10.

Box 2.14
Dreams

Make a list of your expectations when it comes to graduate school and rank them. Discuss the most important ones with your advisor. Doing so allows your advisor to expose you to relevant experiences. Do not expect your advisor to remember all of your dreams. However, if you are part of a large group, you may want to discuss your hopes and expectations often. The shorter your advisor's memory, the more often you can change your mind!

FROM COLLABORATORS

In chapter 9, we will discuss how the kickoff meeting may be used to share specific expectations from all stakeholders. Generally speaking, two themes arise: communication and confidentiality.

The crux of the matter in trying to realize a complex collaborative project is communicating regarding:

progress: the team must be informed of successful tasks and milestones; or

lack of progress: keeping the team in the loop when progress is slower or halted allows everyone to contribute possible solutions and keeps members feeling involved in the project.

With all collaborators, agree on a communication method that suits everyone, and try to stick to the plan.

Research in engineering often leads to the creation of IP (see chapter 10), the prospect of publication in high-impact journals, or is sometimes highly sensitive. As such, some research projects invite more discretion than others. Discuss with your advisor and collaborators what level of confidentiality is expected of you before disclosing ideas and results outside of your group. Collaborators often ask that you sign a nondisclosure agreement (NDA).

Definition 2.12: Nondisclosure Agreement   *The NDA, also known as a confidentiality agreement, is a legal contract between two or more parties that specifies the nature of the confidential material that each party wishes to protect.*

Remark 2.7: Signature   *Whenever you are asked to sign a document, make sure you are comfortable with its content. Ask someone from your school's research office to explain and review such documents with you. Universities typically have one or more lawyers either on staff or on retainer, who are involved in the review of any legal agreements, and with whom one can consult if asked to sign something by an outside entity to ensure it has been properly reviewed.*

3

JOURNEY AND MILESTONES

"Life's a journey, not a destination."

> attributed to Ralph Waldo Emerson,
> popularized by the American rock band Aerosmith

Historically, a milestone consisted of a stone obelisk placed alongside a Roman road to indicate distance from or to a city and reassure travelers that they were on the right track. In modern times, milestones are metaphorically used in project management to indicate significant progress in the stage of development.

Box 3.1
Doctoral Milestones—Step 1

What are the doctoral milestones required by your institution? Which ones are etched in stone (such as the doctoral dissertation), and which ones, if any, are optional?

Figure 3.1 illustrates the typical journey of a PhD candidate from admission into the program to graduation.[1] Common milestones include:[2]

- finishing course work;
- finding a research topic and an advisor;

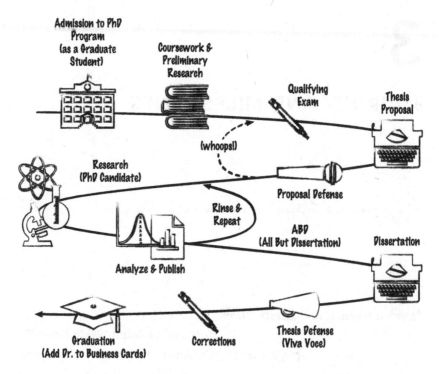

3.1 The journey and milestones. After qualifying, the graduate student becomes a PhD candidate, and, after a successful thesis defense and dissertation, a doctor. A single failure does not imply the journey's end but an opportunity to build stronger foundations.

- passing the comprehensive doctoral exam:[3]
  - qualifying exam;
  - thesis proposal;
- performing research (a journey more than a milestone);
- publishing original contributions;
- submitting dissertation; and
- defending and correcting[4] thesis and wrapping things up.

In chapter 8, we define milestones in the context of project management of a doctoral research project. Here, we introduce the typical journey and milestones of a PhD candidate. The doctoral journey of a candidate, who is initially only a user of knowledge and technology, includes the following phases:

initial **segment:** finishing courses, sketching ideas for a project, getting to
   know your advisor and lab culture;

exploration: situating the limits of knowledge, discovering triggers, drafting a strategy, uncovering ethical considerations;

doctoral exam: assessing of knowledge (breadth and depth), defending a
   research proposal;

research: acquiring results, adjusting the objectives and scope of the
   project;

diffusion: presenting work at a conference, publishing some articles; and

maturation: improving originality, impact, and autonomy.

At the finish line, the student has metamorphosed into a creator of knowledge. The thesis is ready for defense, some more papers are published, and
the candidate may aspire to an exciting career.

Box 3.2
Doctoral Milestones—Step 2

What milestone of the PhD journey do you fear the most? For many people, it
is the (proximal) doctoral exam. For others, it is the (more distant) dissertation
write-up. A common strategy to alleviate fear is gathering information—this
chapter should help you tame the doctoral milestones.

3.1 COURSES, WORKSHOPS, AND SEMINARS

Typical PhD programs require a combination of compulsory and elective
courses, but extremes exist, ranging from research only to a military-like
regimen of prescribed activities. As a graduate student, you are responsible
for understanding the rules of your PhD program and planning accordingly. You will be asked to submit a study plan often agreed on by your
advisor. Plan your coursework to the best of your knowledge, but remember that, contrary to the Roman milestones, such planning is never etched
in stone: students are allowed reasonable changes as they discover new
topics to explore.[5]

Figure 3.2 shows a study plan that you may draft prior to completing the official form from your institution. It shows topics, number of

Study Plan

Mandatory Workshops	For Credit Topics	Advanced Topic
Workshop 1 (1CR) — Y1 (F) Workshop 2 (1CR) — Y1 (W)	Topic 1 (3CR) — Y1 (F) Topic 2 (3CR) — Y2 (F) Topic 3 (3CR) — Y2 (W)	Machine Learning — Y2 (W)
Comprehensive Exam (Preparation) Course 1 (3CR) — Y1 (F) Course 2 (3CR) — Y1 (W)	Optional Topics Intro. to Python (3CR) — Y1 (F) Charm School (1CR) — IAP	Practical/Professional Skills Lasers for Newbies — Y3 (S)

I plan to learn about quantum mechanics, AI, and take Spanish lessons.

But I need you in the lab!

Legend:

F — Fall
W — Winter
S — Summer

PhD Program

Offered Elsewhere

Conferences & Summer School

3.2 Study plans should include a list of mandatory and optional topics, a timeline, and the number of credits for each activity to ensure the school's requirement is met. For the record, Charm School was not mandatory during my PhD, but it was certainly an option offered each year during IAP.[6]

credits, and a tentative timeline: the proper ingredients to prompt a fruitful discussion with your advisor.

COURSES

Courses offer a great way of studying a particular topic and understanding complex theoretical concepts. They are meant to give you the well-rounded education expected from an expert in the field and not explain all the details of your specific doctoral project. Therefore, you should choose them wisely and plan on doing relatively well in them as students are often required to maintain a minimum grade point average (GPA) to remain in the doctoral program. When preparing a study plan, ask your advisor for input on the topic, and make sure to consult fellow students regarding the workload and teaching style. Aim at the most relevant courses and seek inspiring teachers. Inquire about the possibility of attending classes in a neighboring institution, such as a medical or a business school. Being an active participant in a classroom is a great way to meet with potential thesis committee members. Remember, however, that nobody gets a PhD from coursework alone. Perhaps, now is an excellent time to review the 90/90 rule, as the time you spend in a classroom competes with that spent for research.

Definition 3.1: The 90/90 Rule   *According to Tom Cargill, a computer scientist at Bell Labs, the time required to complete the first 90 percent of a software development equals that required to complete the last 10 percent: The first 90 percent of the code accounts for the first 90 percent of the development time. The remaining 10 percent of the code also accounts for the another 90 percent of the development time.[7] In other words, the output of a project does not increase linearly with effort but rather follows an S-shape. I will not go as far as to say that this applies exactly to studying for a class, but, remember, before you spend eighty hours on a single class assignment, that, perhaps, you can stop before it is perfect[8] and spend the remaining hours in the lab.*

Box 3.3
Coursework

What are your criteria for choosing which courses to take? Do you think both you and your advisor agree on your course selection? Do you need to seek classes or workshops outside your institution? Use these as talking points with your advisor when getting your study plan approved.

WORKSHOPS

Workshops are typically aimed at acquiring specific skill sets. They can be about complementary skills or, at the opposite end, very technically oriented. Unbeknownst to many new students are summer schools: a short period packed with lectures from leaders in a field. Such events are a great opportunity to jumpstart your career and build your professional network. Similarly, the Gordon Research Conferences®[9] allow junior researchers to learn and mingle with experts on very narrow topics. Organizers sometimes provide travel grants that could be combined with funding from your advisor. You should also inquire with the organizing committee about opportunities to get involved in exchange for reduced or even waived fees.

SEMINARS

For graduate students, departmental seminars are often synonymous with free food. They, however, have more to offer than compensation for poor stipends: seminars are a great way to broaden your scientific horizons and professional network. That being said, not all seminars are created equal, and I vividly remember how challenging it was to stay engaged—or even awake—through some of them. As a mid-career scientist, I have spent an equal amount of time on either side of the podium. Here are my two cents about these two realities. As a student, I have found it extremely painful to listen to a talk unprepared. However, I found that if I managed to read a paper or two related to the topic before the presentation, the fifty-minute speech became bearable, even pleasant. When I did not manage to read ahead, or if the subject did not offer much in line with my interests, I would try to analyze the presentation itself: Did I like the transitions? The animations? The choice of fonts? I would make notes to incorporate (or not!) such elements in my presentations. As a speaker, I try my best to maintain the audience's interest and have progressed from fearing to embracing the question period, especially when questions are asked from junior audience members. Indeed, by asking a question, you create a rapport with the guest speaker and distinguish yourself from the crowd.

On asking questions   Asking a question is an art. Most senior faculty take it for granted; some even manage to ask the most pertinent question after having napped for most of the talk. Perhaps as an attempt to avoid the *senior* epithet, I make a point to remember how intimidating asking a question is. For your first attempt, I humbly suggest following this recipe. First, break the ice with a compliment (on the talk or the research, never on the outfit, it goes without saying). It does not have to be elaborate—a simple *thank you for your excellent talk* will do. This first step is essential as it creates a rapport between you and the speaker by showing empathy for this person, alone, behind the podium, revealing their work in front of a crowd. Next, ask your question. If the question refers to a particular point in the presentation, ask to go back to the proper slide (see figure 3.3). It is totally fine to read from your notes—remember that even seasoned TV anchors write questions down when interviewing a guest. If you can, avoid multiple-part questions—this too will distinguish you from *senior*

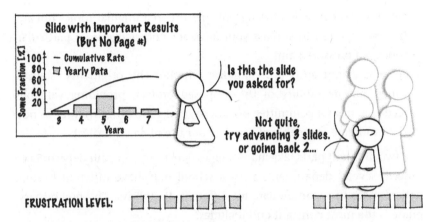

Not quite,
try advancing 3 slides.
or going back 2...

3.3 Asking a question often requires going back to a particular point in the presentation. To avoid the *would-you-go-back-3-slides-no-forward-one-I-meant-one-more-yes-thank-you* waltz, please adopt the habit of numbering your slides!

people asking six questions at once. Ask your most important point first, and if you have additional questions, consider meeting with the speaker during the break.

3.2   THE (DREADED) COMPREHENSIVE EXAM

In most doctoral programs, students are admitted as graduate students, and only upon passing the doctoral exam (or comprehensive exam) do they become PhD candidates.[10]

Box 3.4

Comprehensive Examination

What are, in your opinion, the objectives of the comprehensive examination?

A comprehensive exam typically consists of two parts:

a qualifying part: aims at challenging your knowledge's breadth and depth by validating your understanding of the bases of a discipline and a specific topic. Both depth and breadth are important. Indeed, a PhD holder is often expected to explain topics related to their thesis while becoming the most knowledgeable person on a singular topic. You are indeed expected to surpass your advisor's knowledge on a particular topic. No pressure!; and

a preliminary part: aims at evaluating whether the originality, significance, and feasability of the proposed project, providing valuable feedback from a committee of experts, and validating whether or not the candidate is capable of carrying out the research project.

The two different parts take many shapes depending on your department. Indeed, several departments with a school may have different examination styles to accommodate for different disciplines. The preliminary phase is the most constant and includes:

- preparing a thesis proposal; and
- defending a thesis proposal in front of a jury, which may or may not include your research advisor.

The qualifying phase has the most variability, with the following being a non-exhaustive list of elements:

- obtaining As in several courses;
- writing a general examination on several undergraduate topics;
- taking a specific examination on one or several graduate topics;
- preparing an essay based on seminal papers in the field; and
- answering an oral exam on topics related to the research topic.

Box 3.5
Your Comprehensive Exam

Find out what the rules are regarding the comprehensive exam in your department, including deadlines. Meet with students in your department who have just passed it (and perhaps some who have failed as well) to gauge the scope of your preparation.

How will you prepare for the distinct phases of your doctoral exam? How will you study? Do you have access to exams from previous years? Is the thesis proposal defense public, and, if so, can you attend one to understand the expectations better?

Your department sets the rules regarding the exam timing, that is, how many semesters are you allowed in graduate school before completing the process. Your school also sets the rules regarding failing the exam: which part you are allowed to try more than once. Failing once happens— some of the best researchers I know were given a second chance at one point or another of their career. However, failing twice is a sign that perhaps you should consider other paths to contribute to the research and development ecosystem.

Remark 3.1: Sooner Rather than Later *If you are to fail at getting a PhD, it is better to fail sooner rather than later. The committee does not want you to waste time and resources if it is not confident that you have all of what is required to succeed.*

3.3 THESIS PROPOSAL

In engineering, the thesis proposal typically consists of a written document submitted to your thesis committee for evaluation and in preparation for an oral defense. A thesis proposal[11] is not a mere school or laboratory report but a document for decision. It must revolve around a strong thesis statement, the central point justifying all other document sections. A good thesis proposal (discussed in chapter 4) aims, in no particular order, to:[12]

- define a thesis statement (question, hypothesis, and objectives);
- evaluate your analytical, problem-solving, and critical thinking abilities;
- assess whether or not you are capable of conducting original research independently;
- demonstrate that you have the qualifications (sufficient breadth and in-depth knowledge of the discipline) relevant to the proposed topic;
- validate the originality, significance, and impact of the proposed work;
- verify that you have all required resources (supervision, equipment, money, and time) to lead this project to completion; and
- test how well you can communicate your knowledge of the field.

ON PREPARING YOUR THESIS PROPOSAL

The typical structure of a thesis proposal is discussed in chapter 4. In the meantime, let me say this:

- Every institution has its own rules regarding the thesis proposal: make sure you make yourself familiar with and respect them.
- The thesis proposal defense is not a competition: you are not fighting with others for a spot in the doctoral program but are trying to show that you can conduct doctoral research. You will benefit from asking for help from fellow students, and you will also benefit from helping others, as long as your project is not confidential. If so, make sure that colleagues you enroll are within the confidentiality circle: a more senior student in your group, a postdoctoral fellow, or perhaps a research associate.
- The thesis statement (comprising the research question, the hypothesis, the general and specific objectives) evolves as the literature search progresses.
- The write-up should begin with a strong outline defended to your advisor before filling pages after pages. Reviewing and editing an argument that fits within a single page is much easier (and faster) than to shuffling paragraphs of text spread over twenty-five pages.

Box 3.6

On Scheduling Committee Meetings

Keep in mind that **everyone** tries to schedule a doctoral exam or defense right before the end of a term (i.e., April, August, and December). As a result, schedules are full, and professors are over-solicited. Aiming to meet at various points in the school year will increase the odds of successfully getting the experts you want and might improve their general mood and mental disposition for constructive feedback. Scheduling surveys help converge on a date but avoid alienating your committee members with too many options (see figure 3.4), and provide an end date for a rapid confirmation as it is impossible to pencil in so many options. A helpful hint would begin with checking the teaching schedule of committee members and avoiding periods when you know they are in class.

ON DEFENDING YOUR THESIS PROPOSAL

Entire books are written about oral presentations,[13] and I encourage students to pick one up at least once in their careers. Here, I wish to

Please fill this convenient online survey to identify a time for my Thesis Proposal Defense

3.4 Discuss the composition of your thesis proposal committee with your advisor early, as the most interesting people are often also the busiest. Speaking of busy people, try not to alienate committee members by asking them to fill out a never-ending scheduling survey.

emphasize tips about presenting and defending your thesis proposal. Make sure you:

- inquire about the format: How long should the presentation be? Is there an institutional template? What sections are expected?;
- practice, practice, practice: first, alone, at home, perhaps in front of a mirror, out loud,[1] and with a timer. Then, in front of fellow students, your research group, or students who recently presented their thesis proposal. Once again, respect the confidentiality of your project;
- stick to the rule of thumb for timing: plan for no more than one slide per minute, and try to convey only one important idea per slide. Curate your content to convey the essential ideas for the thesis committee to understand the research you are proposing, its impact, and its significance. And avoid the situation depicted in figure 3.5; and
- prepare for questions: ask your practice audience to grill you with their questions and to anticipate one or two important topics for which you prepare backup slides.

Remark 3.2: Practice Makes Perfect *Practicing out loud allows identifying words you stumble on, or word combinations that are tongue twisters. If you find one, you may elect to rephrase, unless you are indeed studying the **big, blue bug that bled black blood**. In such a case, you have no other option than to practice, practice, and practice some more.*

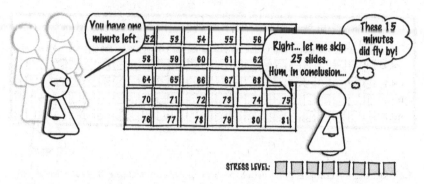

3.5 The rule of thumb for timing a scientific presentation is to prepare one slide per minute. The idea is not to cram as much information as possible but to carefully curate what you want your audience to take home. Once you get a few presentations under your belt, you may consider bending the rule. Slightly.

Box 3.7

Sophomore-Year Physics

When I was preparing my thesis proposal defense, a senior scientist working in the lab said to me: *stick to sophomore-year physics, but be prepared to answer questions at the PhD level.* The purpose of the exam is to test the extent of your knowledge: don't offer all that you have upfront, but make sure you can also entertain tough questions.

ON MASTERING THE SCIENTIFIC LANGUAGE

Common feedback I hear from students goes along the lines of *English is my second (third, fourth...) language—I fear not writing or speaking it well enough to present my work adequately.* This comment is so common that I felt compelled to address it here. Stephen King[14] suggests two ways to improve one's writing skills: to read and write as much as you can. With this in mind, I recommend the following exercise. Every day, read (at least) one scientific paper. For each, write a summary in the note section of your bibliographic database as a writing exercise. In addition, identify which sentence in the paper is your favorite. Keep a diary of these favorite lines—along with their origin—in a separate document. This document will become your custom-built phrasebook. This list will serve as an inspiration for the structure of the sentences in your manuscript. Be careful, however, to only reproduce the construct of the sentence and not copy and paste it word for word.

Box 3.8
Plagiarism

In the previous example, I insist that you only reproduce the *structure of a sentence and not the actual words*. Although some say that gathering inspiration from many authors is research, copying one author is still very much considered plagiarism.

The same idea holds for oral presentations. Whenever you are attending a seminar, watching a Technology, Entertainment, Design (TED) Talk, or even during lectures, write down sentences that you like. While a scientist's dream is to, someday, naturally discuss science in front of any audience, there is nothing wrong with preparing a text for your first lecture, as long as you put in the hours to practice, practice, and practice your delivery ahead of the big day.

3.4 RESEARCH PROJECT

Which one comes first: Finding a research project or a thesis advisor? The two are entangled, especially in engineering, as projects require sophisticated infrastructure only owned by a few researchers in the world. Given the number of resources necessary to perform experimental research, one could argue that what comes first is, indeed, money. According to Larivière,[15] among the top factors contributing to the successful completion of a PhD is receiving a stipend. Interestingly enough, success does not correlate with the amount of money received, as long as the student gets enough to cover basic needs, defined as Maslow's pyramid's first stages,[16] plus high-speed internet. Other success factors include a good relationship with one's advisor and a keen interest in the topic. In this section, we first describe where research money comes from, then explain what to look for in an advisor, and finally, discuss how to define an exciting research project.

FUNDING A RESEARCH PROGRAM

Few students realize that researchers are not given money to perform research but that they must constantly be writing grant proposals to funding agencies. Like scientific publications, research proposals are peer reviewed, but unlike scientific publications, funding agencies have strict deadlines indirectly defining a grant season. For the sake of conciseness, we will distinguish two broad categories of funding sources:

project-centric: researchers submit grant proposals to government agencies, private foundations, or industries to fund research equipment, consumables, and salaries and stipends for research staff, including students. Research proposals must be submitted sometime before recruiting students and staff as the peer-review process is somewhat lengthy. Industry-sponsored research includes research projects (guarantee of effort) and research contracts (guarantee of results). A student paid through their advisor's research money is often said to receive an RA;

scholar-centric: students, postdoctoral fellows, and researchers may apply to scholarships and fellowships to cover their expenses while they join a research team. Scholarship recipients have greater freedom when choosing a research group as their salary does not have to be drawn from the researcher's grant. Scholarships and fellowships typically involve a project description, implying that the student has chosen a research lab *a priori*.

Box 3.9
Funding

Are you already involved with a research lab? Do you know how your research is funded? If it comes from a grant, perhaps you could ask your advisor to read the research portion of the proposal as a way to jump start your literature search. Another great way to learn the ropes is to help your advisor in writing their next round of grant applications.

DEFINING A RESEARCH PROJECT

In his bestselling book *How to Write a Thesis,* Umberto Eco discusses the four rules for choosing a research topic:[17]

the topic must match the candidate's interests and background: for example, a project on elementary particles is more suited for a candidate in engineering physics than for an industrial engineer;

the (re)sources[18] must be available to the candidate: when working with elementary particles, it helps to have access to a particle accelerator;

these (re)sources must be usable by the candidate: following our example, the candidate must know (or be learning) skills required to interpret data from the particle accelerator; and

the methodology must be within reach of the candidate: a student fascinated by equations will thrive more when working on a theoretical problem than when forced to assemble an experimental apparatus, and vice versa.

To these four rules, Prof. Eco adds a final one: the advisor must be the right person to advise the candidate on such a topic. These rules seem rather obvious—again, it goes without saying—but much of the academic life (from applying to scholarships, or grants, to defending a research proposal) consists in convincing a committee of experts that you have the required expertise and resources (e.g., mentorship and equipment) to carry out a particular project.

Box 3.10
Research Question

Once the research field is established, where does one find a good first research question? A sure way to kill a few birds with a single stone consists in reading theses from previous lab members and, perhaps, that of your advisor. Many scientific questions are to be found in the future work section. Other avenues include reading grant proposals from your lab, in addition to surveying recent literature.

FINDING AN ADVISOR

There is no perfect recipe to finding a thesis advisor. For starters, schools have different enrollment policies. In some schools, graduate students are admitted into the PhD program without an advisor. They spend their first semester on classes and laboratory rotation to eventually commit to one laboratory, and thus one advisor. In other schools, you are hired by the laboratory and then become a student. Unless you pursue your graduate studies in the same school as where you spent your formative years, the approach and interview are virtually made through emails and video conferencing.

I have experienced both approaches: I did my PhD in a school that permitted two semesters of exploration before committing to a laboratory; I currently work in a school where advisors are chosen before students are admitted to the school. Both systems have their pros and cons. In my school, students begin their research projects right away. In my previous school, students were given the option—but never a guarantee—to find a perfect match. Different approaches accommodate different funding philosophies. Starting with research from day one provides funding in the form of an RA. When rotating through different labs is encouraged, the school may provide funding through a TA or scholarship. A student who is independently funded through external scholarship has more flexibility.

Box 3.11
Prof. Perfect?

Take a moment to describe the qualities of the perfect advisor for you. It may help to explore extreme, yet common, traits described by Snieder and Larner in their book *The Art of Being a Scientist*:[19]

- the over-committed: is rarely available for you;
- the magnum opus writer: keeps insisting on one more article;
- the critic: points out every mistake without offering fixes;
- the copy editor: critiques the form—from Oxford commas to unbreakable spaces—but never the content;
- the (indifferent) cool cat: is fine with everything, always;
- the hyperactive: has a new weekly priority every day;
- the conservative: stays within their comfort zone;
- the advisor-splainer: holds on to the mike and never listens;
- the competitor: keeps their cards close to their chest, fearing the day you will start your own lab and compete for the same limited pool of funding.

Snieder and Larner[20] conclude that the professional advisor "has ample time for you, reads your manuscripts promptly, gives adequate comments, motivates you, and combines the roles of coach and evaluator admirably, and is thrilled to see your ideas forge ahead of hers."

3.5 PUBLICATIONS

In chapter 2, we described the peer-reviewed publication process. Here, we describe motivations for publishing.

ON THE IMPORTANCE OF PUBLISHING

Communicating research results is the last step of the scientific method. Without publishing, and regardless of how spectacular your research results, the circle is incomplete: it is as if research had never occurred.

Researchers also have a moral obligation to publish. So-called by many[21] history's biggest fraud, the Agricultural Revolution—which increased the *sum total of food at the disposal of humankind*[22]—did not translate into more leisure for the farmer but created both a population explosion and a pampered elite. As a researcher, you are part of the pampered elite. Perhaps this is not obvious to you as a graduate student, but while you are working on fun projects in the lab, someone else is producing food for you. In return, you have a moral obligation to deliver research in a form that is useful to humankind: conference presentations, proceedings, and peer-reviewed articles. While not perfect, peer review is the most reliable mechanism to authenticate studies and contribute to the global database we call scientific knowledge. And since farmers will not wait until the end of your PhD to put food on your table, when possible, try not to wait for the end to disclose results. Publish as you go.

Publishing also follows a pay-it-forward model. The students who published before you allowed your advisor to be competitive when applying for research funding and, ultimately, hire you. The results you publish as a graduate student will contribute to enhancing the grant application for the next grant cycle and help the next generation of students.

Another important reason for publishing is to establish your reputation as a researcher. No matter how competent you are in manipulating scientific equipment or how knowledgeable you are on various topics, you, as a researcher, are evaluated by your publications. Scientists mostly become known (or remain unknown) by their publications.[23]

> **Box 3.12**
> **First Publication**
>
> Have you already participated in the preparation of a scientific article? What are the main difficulties you encountered? If you haven't, ask this question to a more senior student in your lab or graduate student cohort. How can you alleviate these difficulties?

PUBLICATION ROAD MAP

Your advisor has been a researcher for longer than you have and indeed feels the constant pressure of being evaluated with publications. Therefore, even if you have not explicitly prepared a plan for publishing your work, the chances are that your advisor has at least some vague notion of how to disseminate your work.

> **Box 3.13**
> **Strategy**
>
> Take a moment to devise a publication strategy. What would you like to publish? Which journal are you targeting? And to which conference should you present your work—aside from the self-evident choice of attending a workshop in Hawaii? Refine your publication strategy with a senior member of the lab, and then share it with your advisor. Do you both share the same vision? Can you agree on a preliminary road map?

A publication road map contains the following elements:

What to publish: What is the nature of the scientific contribution?

How to publish: Are you considering a short letter-type or full-length article, a conference presentation with proceedings, a review paper, or a white paper?

Where to publish: Select a journal for publication first and foremost on its readership—what journals did you read the most to build up your foundational knowledge? Will your work be helpful to the journal's research community?

With whom to publish: Who are the tentative authors of the paper?

In your publication road map, try to strike the right balance between your dream paper, which may never end up being submitted, and salami slicing.[24] Your publication and research plans should be prepared in parallel: as you decide on the long-term goals of your research, envision such experiments or calculations as a publication or as a series of manuscripts.[25] Writing may begin before data analysis is completed. At the very least, you should reflect on the journal and start familiarizing yourself with its style (by reading more papers from that source) and its template (which you can readily download and adapt). As your research evolves, so will your publication road map.

Box 3.14
Amendments to the Publication Road Map

As the opportunity to discuss my favorite topic—that is, myself—arises, I share in figure 3.6 the evolution of my publication road map from graduate school.[26] My advisor began with a topic: developing a new clinical microscope. This required developing a new laser[27] in collaboration with an uber-talented postdoctoral fellow before performing the first demonstration of the rapid microscope, for which I would be the lead scientist (i.e., the first author).[28] Next, a new probe was to be attached to the microscope system (conference presentation)[29] before performing a first pilot study in the operating room (full-length article).[30] Every contribution raised more questions: Are there other configurations for the laser?,[31] What else can the microscope do?,[32] Should we develop better optical fibers?, Could other medical specialties benefit from the microscope?[33] Some avenues were explored during my PhD and others after I left the lab and began my career as an independent investigator. Maps are not static anymore, nor should your publication road map be—revisit it often with your advisor.

ON AUTHORSHIP

According to the The Institute of Electrical and Electronics Engineers (IEEE), authorship is reserved for people meeting all three of the following criteria:[34]

- having made a significant intellectual contribution to the theory, experimental design, collection, and analysis of data;
- having contributed to drafting and critically reviewing the article; and
- having approved the final version of the article.

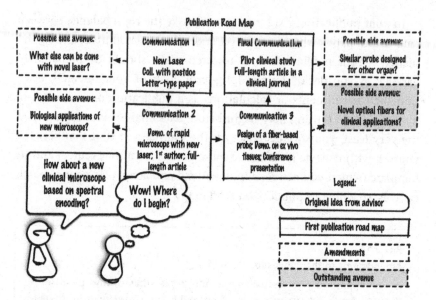

3.6 The publication road map from the author's own doctoral thesis. The speech balloon expresses the original seed idea from my advisor. The solid lines show the original publication road map, while the dashed lines show amendments. The shaded areas highlight the outstanding avenue that turned into start-up funding for my new lab.

Contributors who do not meet all three criteria must be simply acknowledged.

Naming order   Now that you have a list of people: Does the naming order matter? It does. In most engineering fields, the first author is the person who performed most of the work, that is, the person who spent the most hours in the lab assembling the system, countless nights crunching data, and, finally, produced sweat and tears while completing the first draft.[35] The last name on a paper, the senior author, is typically the head of the lab.[36] Everyone else—other graduate students, postdoctoral fellows, or very dedicated interns—is scrambled in between. Deciding early on a list (and ranking) of authors is a good idea in a collaboration involving two graduate students and two senior researchers. Many journals allow two first authors, with an asterisk indicating that the first two names on a paper are co–first authors. This last point indeed illustrates how important publications have become in building the careers of rookie scientists.

3.6   DISSERTATION AND DEFENSE

The dissertation and defense are the capstones of the graduate education and research experience.

THE DOCTORAL DISSERTATION

Along with the thesis defense, the doctoral dissertation constitutes a rite of passage and a credential for admission into the research community.[37] As discussed in chapter 10, it may take different forms:

traditional thesis: written *de novo* in a continuous fashion from start to finish; or

cumulative (or article-based) thesis: presented as a collection of manuscripts (published or submitted for publication) surrounded by original connecting chapters enhancing the cohesion of the ensemble.

Once your dissertation is submitted to your department, it will be evaluated by a jury. The composition of your jury will depend on rules from your school but typically comprises your research advisor and co-advisor, if applicable, an independent external member (where academic independence is defined in definition 2.9), and other scholars in your discipline, including one acting as the president of the jury and one representing the dean. The thesis committee needs not to have the same composition as your thesis proposal jury—again, your school sets the rules and requires approval of your choice of committee members.

TIMELINE

Between submitting your thesis and defending it, two events need to happen. Firstly, you must allow the jury to read and evaluate your dissertation and agree that it is worthy of a defense. This step is out of your control and may take one to a few months. Secondly, your thesis defense needs to be advertised for some time, ranging from several days to a few weeks. This is the perfect time to prepare and practice your thesis defense.

THESIS DEFENSE

The thesis defense is a formal event in which a PhD candidate exposes their research contribution in front of a jury of experts. Its exact protocol is dictated by the institution, with varying degrees of formality. The best way to familiarize yourself is to attend a few in your department.[38] As a bonus, you will learn great science and acquaint yourself with the questioning style of several faculty members. You may even practice asking a question to the PhD candidate. More on this in chapter 10.

COMMENCEMENT AND HOODING CEREMONY

Doctoral diplomas are awarded during the commencement ceremony (or mailed to you if you opt not to travel back to campus for the occasion). Some schools hold an additional celebration for their newly minted PhDs called the hooding ceremony. During this ceremony, a hood will be placed over your head to recognize the few who have completed the highest academic program.

4

THE THESIS PROPOSAL

"*You are not Proust. Do not write long sentences.*"

Umberto Eco[1]

Part of teaching doctoral workshops at PolyMtl involves reviewing the first draft of each student's thesis proposal. One of the general comments I write the most on the margin is **help me help you**. Indeed, most students do not realize *a priori* that the thesis proposal is not just a hurdle sadistic doctoral program directors throw in the way of rookie students. It is, in fact, the first serious discussion you and your committee have regarding your plans for doctoral research. Its purpose is to help the committee help you verify that all boxes are checked before you begin your doctoral journey.[2]

4.1 PURPOSE

The doctoral thesis proposal serves several purposes:

- to verify that your proposed research is novel, significant, and impactful by leveraging the expertise of your thesis proposal committee in addition to your research;
- to assess whether or not you have the depth and breadth of knowledge, enough time, and resources to pursue a PhD on the topic;

- to verify your capabilities to defend and position yourself as an aspiring leader in your research field; and
- to showcase your preliminary data if you have some (not mandatory).

SECTIONS OF A THESIS PROPOSAL

Every school has its preference when it comes to formatting a thesis proposal. Typically, students must respect the provided template. Indeed, your thesis proposal committee members are very busy people: a familiar format ensures that they find relevant information quickly. Figure 4.2 highlights the pretty general structure employed at PolyMtl:

title: a working title that is informative and concise, yet not etched in stone;

abstract: a summary of each section of the manuscript;

introduction: the context and motivation for your research that converges on the research question;

literature review: a synthesis of the state of the art and concepts necessary for your work that converges on gaps in knowledge that will form the basis of your general objective;

thesis statement: the core of your proposal, where you reiterate the research question, state your literature-backed hypothesis, and expand your general objective into specific objectives (SO)s;

methodology: a detailed explanation of how you will perform your work, what resources are required, and what assumptions are made. For clarity, you may group aspects of the methodology according to specific objective (SO)s;

results: a report on preliminary results (if any) and a description of anticipated results;

conclusion: a discussion on the significance and impact of your preliminary results and an outlook on the short-, mid-, and long-term potential impacts of your research;

project management: a timeline of the critical tasks and a list of resources required, and associated risks and contingencies; and

references/bibliography some schools prefer that you indicate your references as footnotes, but most ask for a separate bibliography section.

Thesis Proposal

I propose to
In this proposal, I wish to
Since the early days of

INSPIRATION LEVEL: (NEGATIVE) ☐ ☐ ☐ ☐ ☐ ☐ ☐

4.1 Starting with an outline is a great strategy to fight leucoselophobia (from leukos, selida, and phobos, respectively Greek for white, page, and fear), aka writer's block, or the blank-page syndrome.

WRITING SEQUENCE

Writing your first—and, ideally, only—thesis proposal is intimidating. A strategy to fight writer's block (illustrated in figure 4.1) includes following these steps:[3]

1. put together an outline including your research question, hypothesis, and objectives, and defend it to your advisor;
2. sketch figures and tables, including graphs and plots of your preliminary results;
3. prepare figure captions;
4. write sections on methodology and preliminary and anticipated results;
5. discuss preliminary and anticipated data;
6. conclude with the expected impact (see chapter 7) and significance of your future work;
7. assemble key concepts necessary to understand important points from each section into a relevant introduction; and
8. finish with the abstract and consolidate your bibliography.

The literature search does not explicitly appear in the writing sequence. Indeed, it should not be seen as a milestone but a continuous process. Preliminary scouting of the literature is necessary to identify a research question, and a further search is required to clarify your objectives and methodology. You should stay up to date with current research for as long as you plan on being a researcher (more on this in section 4.4).

If you cannot help but insist on beginning with the introduction, stick to its backbone before adding too much flesh. In his book *How to Write a Thesis*, Umberto Eco suggests the following fictitious introduction: "In this work, we propose to show XX. A literature search identified this gap in knowledge or a lack of data. In the first core chapter, we will try to demonstrate A. In the second core chapter, we will show B, etc."[4]

This helps you establish a narrative for your proposal. Discuss it with lab mates. Only when you are happy with how it flows should you begin to write the introduction section in its final form.

SCIENTIFIC WRITING

Scientific writing sounds intimidating, yet, it is, in many aspects, like most forms of writing. A strong narrative sustains the reader's interest from the first to the last word. Some adjectives enhance the experience; too many adverbs clutter the mind. What is unusual about scientific writing is the topic and the writer. Research results are novel and perhaps complex. Authors are typically novices and often use English as a second language. When advising students, I side with Umberto Eco and suggest that you stick to short sentences.

The complexity of the style should be inversely proportional to that of the topic. Arguably, Proust's feelings for madeleines[5] dipped in tea were complex and, perhaps, justified the five-hundred-word sentences. But if I described nonlinear optics with so many clauses, punctuation marks, and flourishes, no one would make it to my second lecture.

OUTLINE

The first task associated with writing is to prepare (and agree on, when writing with coauthors) a solid outline. An outline is a one-, maybe two-, page document describing critical elements of a manuscript as a series of bullet points, fragments of sentences, and, perhaps—but less ideally so—some complete sentences as well. Why such a compact document? Beginning with an outline allows for rapid feedback from your advisor or fellow lab members. The outline should already follow the required structure for a thesis proposal, with subsection headers adding specificity to your work. The outline serves several purposes:

- making sure every section is addressed;
- organizing your thoughts;
- shuffling ideas without worrying about grammar (yet); and
- viewing the ensemble without having to sort through pages of text.

The outline should highlight sections of your document and contain as little text as possible: no paragraphs, but short sentences or key points. Rely on citations to avoid writing lengthy paragraphs.

Figure 4.2 shows an outline prepared for a doctoral project. The margins showcase tips that I use for scientific writing:

reference placeholders: unless you have all your reference handles at the tip of your fingers, avoid breaking your creative flow by stopping to find a reference. Instead, use a placeholder, such as [REF], that you can easily find later on and replace with the right citation; and

custom bullet point: number your specific objectives, for example, SO1, and use this numbering in the materials and methods (M&M) and in the results sections to indicate what method and results are associated with each SO.

The outline presented in figure 4.2 is fairly detailed—indeed, it is the result of many iterations between the student and me. From the student's point of view, the outline avoids the frustration of reshuffling paragraphs and adjusting the connecting sentences accordingly. From the advisor's point of view, it allows for gaining a global perspective of the argument without reading (or suffering?) through pages of text. #win-win.

Use subtitles to allow your reviewer to find the information faster, and verify that you have addressed all sections.

Use placeholders to avoid breaking your writing flow. Always use the same (e.g., [REF] to easily find them afterwards with control-f).

Bullet points (instead of paragraphs) allow for a comprehensive view of the document within a few pages.

Title - A novel endoscope for the early diagnosis of ovarian cancer

Trigger (Introduction)

- Ovarian cancer kills XX women each year worldwide [REF];
  - Ovarian cancer is a silent killer -> women find out when it is already advanced [REF];
- No screening tool currently exists [REF];

Frontier of knowledge

- Confocal microscopy allows virtual biopsy [REF];
- Confocal microscopy is too large and bulky to image inner organs in vivo;
- Researchers [REF] have shown fibre-based confocal endomicroscopes;
- Current specs are insufficient small tutor detection [REF];

Research Question

Can dedicated fibres improve the size and resolution of confocal endomicroscopes to allow early detection of cancerous lesions of internal organs, such as the ovary?

Hypothesis

Double-clad fibers coupled to spectrally encoded confocal microscopy (SECM) may contribute to improving the diagnostic potential of confocal endomicroscopy.

General Objective and Specific Objectives (SO) ———> Always define acronyms.

Improve specifications (size and contrast) of confocal endomicroscopes through novel fibers and spectral encoding towards early detection of ovarian cancer.

[SO1] Design, pull and characterize a new fibre in a tabletop SECM system;
[SO2] Prototype a miniature SECM and validate it on ex vivo samples;
[SO3] Characterize the SECM probe with novel fibre on a pilot study.

Strategy (Methodology)

Pro-tip: Identify with a custom bullet the methodology and results associated with each objective.

[SO1] Adapt a commercial system to retrofit the novel fibre and scanning scheme, use a standard resolution target (USAF51) and calibrated tissue phantoms. Use signal-to-noise ratio (SNR) as the metric.
[SO2] Simulate ray tracing using Zemax (or equivalent) and opto-mechanics using Solid Works (or equivalent). Agree on appropriate model and validate availability. Compare images with gold standard (histopathology).
[SO3] Depending on the size, per operative or per cutaneous investigation of ovaries.   Not overly specific, but allows planning for availability.

Expected Results, Originality and Impact

[SO1] A novel microscope based on a new illumination/collection scheme. Patent & Publication;
[SO2] A new probe and an original scheme to compare with the gold standard. Conference;
[SO3] Pave the way towards better diagnoses of ovarian cancer. High impact publication.

Anticipated risks and approaches to manage them

- Prohibitive cost to pulling own fibre -> partner with a fiber lab with tower;
- Final probe too large -> can this be used in other organs?

Principal Resource Required

- Ex vivo tissue, partnership with clinician for access to patients

4.2   An outline of a thesis proposal with annotations.

Box 4.1
Peer Review of Your Outline

With a fellow PhD student, share the outline of your respective thesis proposals. Review each other's work and comment on:

- the flow of ideas;
- the coherence of the argumentation; and
- whether sections are missing.

4.2 FRONT MATTER

The front matter of your thesis proposal includes your title page and abstract. Inquire whether your institution provides a template and familiarize yourself with it.

TITLE

The title of your research project should be seen as a progression into the classification of your work. In chapter 7, we discuss the passage from discipline, research field, and topics: your title is what comes next. It should reveal your work's main (expected) result while being clear and concise.

Remark 4.1: *Contrary to distances written on Roman milestones, the title of your research proposal is not etched in stone and will most likely differ from the final title of your dissertation.*

Some tips for choosing a title—for your thesis proposal, dissertation, or journal articles—include:[6]

- emphasizing your main (expected) result as opposed to several topics;
- using fewer than ninety-five characters—papers with short titles (and free of punctuation such as question marks, hyphens, columns, and so on) are on average cited more;[7]
- spelling out acronyms; and
- respecting guidelines from the school or journal.

ABSTRACT

The abstract is a condensed summary of your entire thesis proposal: each proposal section must be represented. Chapter 11 describes the typical structure of an abstract and strategies for its preparation. Since the abstract is a document summary, it is written last.[8]

Remark 4.2: Word Count *Make sure you also consult your institution's guidelines, as there often exist rules regarding language and word count.*

Typically, verbs in past tense describe preliminary research, and verbs in future tense describe proposed work (more on verb tenses in chapter 11).

4.3 INTRODUCTION

In the introduction, you justify the emergence of your research project. Then, you describe the triggers and present the protagonists. You also describe the organization of the core chapters to the reader.

Remark 4.3: *The difference between the introduction and the literature review is that the former indicates why you choose this topic and the latter synthetizes the current state of the art in that field and critiques previous research to highlight gaps in the knowledge.*

TRIGGERS

Which gaps in knowledge are you trying to fill with your research? In engineering, triggers belong to one of three categories:

scientific research: a gap in knowledge;
technology development: an application; and
innovation: a market need.

Chapter 7 further describes the emergence of a research project, where one carefully defines what will be the focus of their professional lives for the next few years. The trigger section sets the context of your research and conveys your enthusiasm for the topic. It describes the current versus the ideal situation.

STAKEHOLDERS

Like opening credits running at the beginning of movies, a small section of your introduction presents the protagonists involved in your project. This is not an acknowledgment section but a paragraph explaining, for example, the collaboration between your lab and another institution (another lab, an industry, hospital, or research center) at the heart of your research project.

ORGANIZATION

It is customary to tell the reader what comes next. The reader, familiar with thesis proposals, indeed expects the results of a literature search, your thesis statement, methodology, and expected results. Try to highlight the specifics of your thesis proposal. Is your project multiply disciplinary? If so, warn the reader that the literature search will begin with discipline 1, followed by discipline 2. Have you already published preliminary work on the topic? Tell the reader that the methodology and results chapters will contain a section on preliminary work, followed by the proposed research. Still unsure how to structure your introduction or any chapter of your thesis proposal? Kindly ask a more senior student to read theirs.

Box 4.2
Protagonists Come and Go

When multiply authored publications are used for a cumulative (manuscript-based) thesis, some schools require that coauthors sign a form certifying what the contributions of the doctoral candidate were. If this is the case for your institution, make a habit of celebrating each manuscript submission by asking each coauthor to sign that form before they leave for a lab far, far away.

4.4 LITERATURE REVIEW

If the introduction section leads to a gap in knowledge, the literature review should convince the reader that the topic is worth pursuing, the question significant, and its solution impactful.

WHY?

As illustrated in figure 2.2, your contribution to knowledge builds on decades of previous research. The literature review aims at situating your research question in the context of what is already known.[9] Its objectives are to:[10]

- organize previous work in the context of how it contributes to your research topic;
- prevent you from duplicating what has already been done;
- highlight relationships and conflicts between previous studies;
- improve your motivation for and understanding of the topic; and
- situate your original work with respect to that of others.

Remark 4.4: A Synthesis *The literature review of your thesis proposal is neither a phone book—a mere listing of previous work—nor a textbook: you should focus on essential results from previous research and not reproduce entire derivations published elsewhere. Cite the work of others, but refrain from using this review as the precursor for a lecture series on the topic.*

The background information provided in your literature begins with a somewhat broad picture and narrows down to your topic, citing relevant work in the process. How far back should you go? Einstein? Maxwell? Newton? Archimedes? Unless you plan to refute Archimedes, a good starting point is midway through the undergraduate degree in your field. Anything more advanced than sophomore-year level should be explained (or reminded) to your readers. If your research is at the intersection of several domains, you may consider separating your literature review into several sections: one for each field. Box 4.3 illustrates the scope of the literature review associated with the project outlined in figure 4.2.

Box 4.3
Literature Review

Figure 4.2 outlines a research project on an imaging instrument applied to a particular pathology—ovarian cancer detection. The literature review should describe the relevant work in microscopy and the pertinent concepts of gynecology. By pertinent, I mean concepts necessary to understand the problem at

---

Box 4.3 (continued)

hand. Try to identify which concepts are relevant to the project and should be cited in the clinical section of the literature review.

1. Ovarian cancer is the gynecologic cancer with the highest mortality.[11]
2. The CA-125 blood markers may indicate the presence of ovarian cancers, albeit with limited sensitivity and specificity.[12]
3. An important fraction of ovarian cancers originate from the Fallopian tubes.[13]
4. The ovary is located in the pelvic cavity, on either side of the uterus.[14]
5. The ovary is responsible for hormonal fluctuations.[15]

Answer: arguably, (1.) belongs to the trigger section, as it motivates the work, (2.) should be part of the literature review as a way to benchmark a new method, (3.) is useful to justify looking at the Fallopian tubes with the future instrument, and (4.) helps to situate the reader and to justify an endoscopic approach. On the contrary, (5.) is too much information, unless one wishes to investigate the role of hormone fluctuations on images acquired with the new instrument.

The scope of the literature review depends on the arguments you are trying to make and should not be used as a platform to show off your erudition. Every article you describe should support at least one aspect of your thesis: the research question, objectives, methodology, results, or discussion.

BACKWARD AND FORWARD IN TIME

A good starting point for a literature search is a seminal publication in your field. From such a starting point, one can move backward by looking at publications referenced in this first paper. Conversely, you can move forward in time by finding papers citing this first article. Another excellent initial document is the doctoral thesis of the student who just graduated from your lab.

KEEP UP WITH THE GOOD WORK

Some engineering fields move extremely fast. Reviewing literature should become a regular habit, as opposed to just an early phase of your PhD. But where do you get started?

PhD theses: unless you are the first PhD student in your lab, consult recent doctoral theses from alumni of your research group—and explore important papers from their bibliography;

seminal papers: from important papers, travel backward in the paper trail to find the seminal paper in your field;

forward thinking: use library databases such as Web of Science or online tools such as Google Scholar to find out who has cited the thesis or the important papers. You may also set up an alert for new citations to important papers in your field to keep yourself up to date with what competing groups are doing; and

tables of contents: subscribe to newsletters from journals publishing the table of contents of their newest edition.

Explore tools to automate your searches: your school's librarians most likely host workshops to improve your navigation skills—consult them!

Automate your work I once almost made a student cry. I had casually asked that he add a reference in the introduction chapter of his master's thesis, only to find out that he had manually created the bibliography. Adding a single reference meant renumbering every item on his list. For reference management, don't be like that student:[16] automate your work. Furthermore, invest in your future self by building a database of articles you will cite in your thesis proposal, reuse for every article you will author, use and improve in your doctoral thesis, and very likely exploit again when writing your first grant proposal.

Definition 4.1: Citation Management Software *Also called a bibliographic software, a citation management software (CMS) imports, stores, and formats metadata (names of authors and journal, the title of the article, date, pages, volume, issue, and so on) to allow you to build a bibliography in your word processor according to the style (author first versus title first, and so on) required by your school (or publisher).*

Not all CMSs are created equal. Some are free while you are a student, and some are free always.[17] Some are stored locally on your computer; some are web-based and stored in the cloud. As I write this book, popular options include:

BibTeX: a free database compatible with LaTeX—it allows formatting a bibliography in a LaTeX document according to a variety of styles, including the most common styles in engineering;

EndNote: a commercial CMS including a database and widgets allowing cite as you type in word processors such as Microsoft Word;

Zotero: a free and open-source reference management software compatible with word processing tools; and

Mendeley: a commercial CMS developed by Elsevier and compatible with word processors.

Conversion between databases is possible, so the choice you make at this stage is not critical as long as you follow these important rules:

- *do* use a CMS; and
- *do not* manually fill your database, but instead, use automation from websites such as Google Scholar,[18] PubMed, and Compendex to fill database entries for consistency.

Outsource (Some of) Your Work When it comes to scouting a new research field for interesting papers, the more the merrier: join a journal club in your lab to hear about articles found by other lab members, and discuss the best pieces you find. Your lab does not have a journal club? Be that great lab citizen who sets one up.

4.5 THESIS STATEMENT

The thesis statement presents the core of your research project through a research question, hypothesis, and general and specific objectives. There is some debate about the need to write a research question and a hypothesis in engineering research. Ultimately, you and your advisor must agree on what to do based on your school's guidelines. However, I will argue in the following pages in favor of including all four horsemen in the thesis statement.

RESEARCH QUESTION

Despite years of research in science and engineering, knowledge gaps, imperfect techniques, and missing products worthy of your input remain. For all three categories, it is possible to formulate some research questions:

Scientific research: How fast is the universe expanding?

Technology development: Can messenger ribonucleic acid (mRNA) technology contribute to creating a vaccine for COVID-19?

Innovation: To what extent does double-clad fiber-based light detection and ranging (LIDAR)s improve the detection accuracy of autonomous cars?

As a PhD student, you will most likely contribute to answering a question provided by your advisor. Notice that asking the question does not suggest the answer, nor does it say how to tackle finding the solution. The research question only highlights the knowledge gap you are attempting to fill.

Your question provides meaning, structure, and direction to your doctoral endeavor. It should be original, important, and finely worded.[19] When formulating your question, look for the following:

clarity: a good question should be easily understood, not able to be interpreted in several ways, and must reveal *a priori* notions (i.e., scientific biases) that should be verified as opposed to wrongly assumed;

concrete: upon reading the question, one should be able to recognize the engineering field;

originality: Does the question naturally arise from previous research, or is it completely groundbreaking? In chapter 9, we discuss risks inherent to research questions in relation to how original they are;

impact: Is the question relevant, nontrivial, and is it important to answer it? Now? It cannot be answered by a "yes" or a "no";

scope: Is it reasonable to assume that the question may be answered by a trainee during the *normal* duration of a PhD?; and

interesting: Is the question interesting enough to sustain your interest (and that of your advisor) for the long run? Does it mobilize your core competencies or competencies you wish to develop?

When is a good time to formulate a research question? As the role of a question is to guide you through your PhD, the sooner, the better. Your question will be refined as you read more and defend it in front of your advisor and lab mates. Remember, your question is only definitive once you defend your thesis. Use a concept map to refine a primitive, perhaps naive, question.

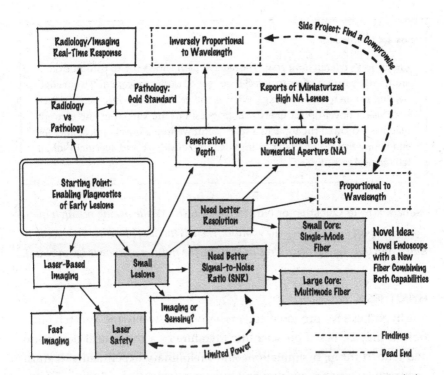

4.3 A concept map based on *a priori* knowledge of biomedical engineering coupled to findings from a thorough literature search on early diagnostic method. The rounded box indicates the starting point. The shaded elements show curated topics that make up the research question. The path labeled as dead ends we revealed by the literature search as already solved. The double arrows highlight opposite behaviors, requiring specifying a compromise value through a side project.

CONCEPT MAP

One strategy to help create a research question is constructing a concept map from the keywords identified in the literature review. To build a concept map, begin with brainstorming concepts and keywords and arrange them graphically to highlight the hierarchy and links between them.

Box 4.4

Research Question in Biomedical Optics

The concept map shown in figure 4.3 explains the rationale behind a research question from a previous graduate student in my group. The starting point is

---

Box 4.4 (continued)

the project's overarching goal. The shaded path shows the transition from a knowledge gap (e.g., How to detect lesions at an earlier time point?) to an engineering research question (e.g., What is the gain in diagnostic performance from using novel optica fibers?). Both questions are valid, but the concept map has narrowed the topic to a specific innovation: a novel optical fiber for endoscopy instead of trying to improve the sensitivity and specificity of all radiological techniques.

Remark 4.5: To Question or Not to Question?   *While asking research question is not part of every subfield's culture, there is merit of phrasing one if only to make sure that your project aims at pushing the frontier of knowledge outward.*

HYPOTHESIS

The hypothesis is a proposed and provisional answer to the research question. It may be based on scientific literature or a theoretical derivation and refined through simulations and preliminary experiments. It must be plausible, testable, precise, and communicable. It should be refutable or contestable and include study variables:

independent variables are standalone measurable quantities that are not affected by other quantities in the study. Someone's age, for instance, is an independent variable, but it cannot be modified. An example of a modifiable independent variable is the location of a tool within a factory; and

dependent variables are measurable quantities that vary in relation to other quantities. The duration of a PhD, for example, is dependent variable as it may change according to countries, discipline, and, perhaps, the age of the candidate. Another example of a dependent variable is the time required to manufacture a product, based on the location of the tools within a factory. One could modulate the independent variable (location of tools) to measure its effect on the dependent variables (manufacturing time).

Box 4.5

Hypothesis in Biomedical Optics

The example in box 4.4 resulted in the research question: What is the gain in diagnostic performance from using a novel (e.g., a double-clad) optical fiber? Key performance metrics in medical imaging performance are sensitivity and sensitivity. Therefore, a testable hypothesis answering the research question may be stated as follows: double-clad fibers with custom specifications [diameters and numerical aperture (NA)s] will increase the sensitivity and specificity of optical imaging applied to oral lesions. Here, the independent variables are the fiber's parameters (diameters and NAs). In contrast, the dependent variables are the sensitivity and specificity of a given imaging technology for detecting oral lesions.

Definition 4.2: Assumption   *In research, an assumption is a statement that is accepted as true to simplify a calculation but provides boundaries within which results are valid. A typical assumption in mechanics is that materials are isotropic, linearly elastic, and homogeneous. Calculations are thus simplified as only the first term of a Taylor expansion is conserved. But results are only valid as long as the assumption is true. Other calculations must be made for anisotropic materials, for example. Make sure you clearly state your assumptions in your thesis statement.*

Remark 4.6: To Hypothesize or Not to Hypothesize?   *While the hypothesis is the essence of experimental research, it is not used in descriptive research in which variables are observed in their natural habitat, without modulating test variables. Discuss with your advisor as to the culture within your subfield of engineering.*

OBJECTIVES

A research proposal—be it a thesis proposal, a grant, or a scholarship application—should describe one general objective and a few specific objectives.

The general objective concerns the global contribution researchers hope to achieve through the research project. One general objective is sufficient—pick the *one to rule them all*.[20]

Specific objectives relate to the activities (read the actions) the research team plans to perform to fulfill the general objective. A PhD project typically includes between three and five specific objectives. The sum of all contributions towards SOs represent the progress made toward reaching the general objective.

Box 4.6

Climbing Mount Everest

Your doctorate can be considered the project of a lifetime. For Sir Edmund Hillary and Sherpa Tenzing Norgay, climbing Mount Everest was a lifetime achievement analogous to others completing a PhD. Their general objective: reaching the summit of Mount Everest (and returning to base camp alive!). I imagine they did not just arrive at the base and began climbing. A more rigorous approach presumably involves defining specific objectives— or actions—required to accomplish such an ambitious goal (i.e., the general objective). I am no climber, but I could imagine breaking their project into the following oversimplified list of SOs:

SO1: get in shape;

SO2: develop and gather the proper equipment;

SO3: map the best route;

SO4: travel to Mount Everest; and

SO5: perform the climb.

Each SO addresses a unique facet of the general objective, and together they form a complete set. Writing such a list allows you to announce the actions you plan to undertake in a way that a reviewing committee may help you identify missing steps. If the ensemble is incomplete, your thesis committee shall assist you in proposing additional measures. However, if you do not submit a list, you make their task much harder and risk them not promptly identifying gaps in the plan. Make sure not to confuse SOs and methodology. For instance, the methodology associated with *getting in shape* could be *train on a smaller summit*.

Box 4.6 shows a list of SOs in the proper form: each item begins with a verb, an appropriate word to describe an action. As general advice:

use active verbs such as validate, characterize, measure, calculate, develop, model, determine, verify; and avoid passive verbs such as understand, study, evaluate, estimate, judge.

Objectives specified by an active verb are more readily quantifiable. Other attributes are memorized through the acrostic SMART: [21]

specific: determines what action you wish to perform;

measurable: provides a way to quantify to what degree the SO is achieved;

achievable: verifies that you possess the appropriate knowledge, skills, and toolset to perform the action;

relevant: ensures that the objective (and the project) is aligned with the mission of your laboratory; and

timely: occurs at an opportune time for the organization and warrants that the action has preliminary endpoints.

Box 4.7
The SMART Way to Climb Mount Everest

Rewrite the list of SOs from box 4.6 to reflect SMART attributes. Not every objective satisfies all qualities, but you will get closer to calling your objectives *specific*.

In the same way that two distinct routes up Mount Everest exist, different researchers may tackle a general objective using massively different SOs. For instance, the general goal: *finding a new cure to a particular type of cancer* may be addressed from the points of view of molecular biologists (i.e., concocting a new drug cocktail), surgeons (i.e., devising a new excision technique), or biomedical engineers (i.e., developing a new laser-based therapy). Given their contrasting skill sets, their SOs are orthogonal. Indeed, I would not let a laser engineer try out a new pharmaceutical cocktail on me (or a new scalpel technique, for that matter!). Box 4.8 illustrates why a complete list of SOs must be included in any research proposal: its reviewers—the thesis committee members—must have enough information to verify that the proposed actions match the expertise of the team.

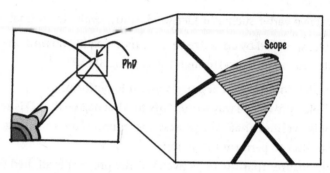

4.4 The scope of a research project shown as the area under the dent made by your PhD to the circle of knowledge.

Box 4.8

Improving Autonomous Cars

Self-driving cars are not fiction anymore, and many groups aim to improve their reliability (general objective). Many students may work on a similar general objective throughout campus without competing with one another. Indeed, the difference lies in their respective list of actions, that is their specific objectives:

computer science student: developing algorithms and databases to improve the computing speed;

computer and electrical engineering students: improving underlying hardware to provide safer and error-safe digital equipment;

engineering physics student: enhancing the reliability of ranging sensors (presumably based on LIDAR) around the car; and

philosophy student: articulating the moral code to solve self-driving cars' dilemmas.

The role of the thesis committee (or grant agency, the equivalent for grown-up researchers) is to make sure that the research team has the required expertise to lead the project to a successful ending.

The scope of a research project is the extent of what you wish to explore. Where does your project begin? The answer is at the frontier between what is known and what is not, as shown in figure 4.4. But, more importantly, where does your project end? Do you stop at some minimum viable product (MVP), proof-of-principle or concept, or do you need to have all i's dotted and all t's crossed? Figure 4.4 shows a depiction of the scope: the area under the dent in the circle of knowledge made by your doctoral contribution.

Box 4.9
Onion Skin or Pilot Study

When describing the objectives of a research project involving a new medical instrument, the starting point is an idea, and the objectives include, for instance, showing a better signal-to-noise ratio (SNR). There are several ways to show an improvement in SNR: A better image of some onion skin? Some cadaveric tissue? Or is the expectation that full translation of the technique be made through a pilot study on humans? When discussing the objectives, also describe the intended scope of your project.

4.6 METHODOLOGY

Since 1991, a satiric prize called the Ig Nobel Prize has been awarded annually for research *that first makes people laugh, and then makes them think*, but for some, it highlights research *that cannot, or should not, be reproduced*.[22] To avoid being invited to the ceremony, you should strive to generate reproducible results. The methodology section aims at providing enough details for another researcher versed in the field to reproduce your work and assess the validity of your results. In a thesis proposal, you describe the methodology used to acquire preliminary results and the strategy to obtain anticipated ones.

PRELIMINARY WORK

The methodology section should describe the M&M deployed to obtain preliminary results. Use shortcuts by referencing commercial equipment by manufacturer and model numbers; custom-made setups from previous work, including the thesis from former lab members; and techniques using publications whenever you can.

Box 4.10

Compact Methodology Section

Here is an uber compact methodology description using part numbers and references [SO1—Demonstrate fluorescence microscopy]: Preliminary data were acquired using a laser diode (L785P5, Thorlabs, Inc., US) centered at 785 nm and coupled to a custom-built spectrally encoded confocal microscopy (SECM) microscope[23] modified to allow for fluorescence detection[24] using a tailored double-clad fiber coupler (DCFC1300L, Castor Optics, Inc., Canada).

Here, despite laser specifications including many parameters, only the wavelength is explicitly mentioned as it is the only relevant parameter in the context. The interested reader is only one Google search away from learning about a specific piece of equipment. When mentioning several specifications is unavoidable, listing them in a table (cross-referenced in the body) avoids saturating your text with numbers.

Remark 4.7: Materials and Methods *This section should serve the narrative of your proposal, as opposed to reproducing its exact chronology. For instance, avoid describing each failed attempt in detail. Instead, relate the latest experiment integrating all considerations learned from previous iterations.*

PROPOSED WORK

The specific objectives describe what you wish to accomplish with your project, while the methodology describes the steps required to reach these objectives. As such, I suggest dividing your methodology section with headers referring to each SO. Document each experiment in the manner of a chef inventing a new recipe. In the context of a thesis proposal, the methodology section allows the thesis committee to assess whether or not the experiment is feasible and the equipment and/or expertise available to you.

4.7   RESULTS

The results section of a thesis proposal presents preliminary results (if any) and describes anticipated results.

PRELIMINARY RESULTS

Ideally, your thesis proposal would present some preliminary results. After all, at the time of submission, you will have been part of the PhD program for around a year. However, you should not worry if you do not have tangible results to show yet, as long as you have traced a convincing path toward novel and significant work.

ANTICIPATED RESULTS

Results should be the pillars supporting (or not) your hypothesis, allow for validating your methodology, and show advances in the field. In this section, you present the results you anticipate for each SO of your project. Visual tools such as a publication road map may help highlight the significance and impact of your anticipated results.

FIGURES, TABLES, AND CAPTIONS

In science and engineering, well-designed figures are worth a thousand words (and several thousand dollars in data acquisition costs). Out of the myriad of visual elements, pick the vehicle that best does justice to your results.

A schematic diagram or flow chart is a drawing used to explain how something works (important parts of an experimental setup, flow of date, a pseudo-code, and so on) with a legend explaining the symbols.

A picture is an image showing what something looks like, with a scale bar indicating dimensions.

A table presents and classifies information (such as results from an extensive literature search or experimental data points), but is seldom used to identify trends.

A scatter plot is a series of data points showing the relationship between one variable and another under varying experimental conditions. It may be used in conjunction with regression curves to validate a fit between experiments and theory. A legend links the symbols to the experimental conditions.

A histogram represents data sorted into groups, useful for studying statistical distributions.

Remark 4.8: Format   *The landscape orientation works best if you plan on reusing your visual material for a presentation. For a publication using a two-column text setting, the portrait orientation is preferred. When preparing a publication, always read the instructions regarding the maximum number of figures. When very few are allowed, consider combining graphical elements into a single composite figure.*

Each visual element must:

- have either a title (numbered and placed above a table) or a caption (numbered and placed below a figure)[25]—use distinct numbering for tables and figures; and
- be referenced in the body of the text, as in "Fig. X shows."

Put some effort into your figures and their captions. An experienced reviewer (or a busy thesis committee member) often reads the abstract, the thesis statement, then evaluates your figures to draw preliminary conclusions. The rest of the exercise consists in verifying whether your interpretation matches theirs.

4.8   DISCUSSION AND CONCLUSION

CRITIQUE

In the discussion section of a scientific document, authors:

- comment on the degree of achievement of each specific objective;
- critique and interpret the results in light of a particular context or application. For example, if the results section states that some parameter was measured to be worth $\alpha \pm \beta\%$, the discussion section comments on the difference between α and the expected value, and whether or not this is usable in the context or whether β could be reduced using an alternate approach; and
- offer insight related to avenues that either did not work or were not explored.

In a thesis proposal, if you have tons of preliminary data, you may discuss these results along the lines just discussed. If you do not have such advanced material, you may discuss the feasibility of each specific objective.

IMPACT AND SIGNIFICANCE

In their concluding statement, the authors summarize their main achievements and emphasize the novelty and significance of their work. You may also provide some outlook on the field. How impactful will these results be for the field, the discipline, and, ultimately, society? Who will benefit from these findings? What are the outstanding issues? In a publication, you also provide insight into what remains to be done, perhaps by you or the next student.

In the context of a thesis proposal, you do not so much look back as you look forward. What is the anticipated short-, medium-, and long-term impact of your work in the field, discipline, and society? The implicit impact on you is that you will get a diploma and, perhaps, become rich and famous, but you must keep this impact implicit and focus on why your work contributes to making this world a better one.

Box 4.11

Engineering Faculty Are Not Archaeologists

Thesis proposals share with scholarship and grant applications that they are evaluated by busy faculty members (pleonasm?)—not all gifted with the same level of archaeological skills. Thesis committee members read your document once and, during the oral examination (or committee discussions for scholarships and grants), must quickly unearth key aspects of your proposal. Do not underestimate the power of a dull sentence such as: *The novelty of the proposed research is* ... to put the spotlight on originality, where even the most absent-minded faculty (pleonasm?) will find it. A more opaque sentence such as: *While other groups have proposed A, we plan on doing A + Δ* leaves the reviewer the burden of calculating Δ, before judging whether Δ is only marginal or, indeed, good enough. A worse proposal would leave the reader extrapolating how original the work is.

Be boring. Tell your reader why you think your work is novel (and significant, and impactful, and ...). Do not hide these features behind elaborate style (or grammatical errors).

4.9 BACK MATTER

At this point, your thesis proposal committee has formed an opinion about the novelty and significance of your work. They now need to evaluate whether or not the project is feasible within the constraints of a PhD, and with the resources to which they have access.

PROJECT MANAGEMENT

Part II of this book discusses strategies for managing a research project in engineering. Chapter 8 describes planning tools such as Gantt charts, work-breakdown structure (WBS), and budget. Chapter 9 further discusses risk management. Here, let me only mention that a good research proposal should include the following:

timeline: including course work, doctoral exam, major experiments, and milestones;

material resources: a list of major pieces of equipment, especially shared pieces or loaners from partners;

human resources: Do you plan on mentoring interns and master's students to increase the scope of your project?;

budget: As the PhD is a stepping stone toward your independent career, do you have some sense of what your research costs?; and

risk management: critical risks and mitigation approaches, when appropriate.

REFERENCES AND BIBLIOGRAPHY

Throughout your proposal, you must cite the work of others or perhaps your own and include a list of such work at the end of your document. First, review the guideline from your institution regarding the format of the references. Then, ask your librarian for advice on building a solid database that will form the basis for your research career. If you have published preliminary data, or have contributed conference presentations, include a section highlighting your talks, posters, or paper as a further indicator that you are on the right path to becoming a prolific scientist.

5

THE AFTER (SCHOOL) LIFE

"The most interesting people I know didn't know at 22 what they wanted to do with their lives. Some of the most interesting 40-year-olds I know still don't."

Mary Schmich[1]

Embarking on your PhD means that you are on a quest toward displacing the frontier of knowledge. It also means you are running out of options to remain a student. In addition to creating new knowledge, you must prepare for the after (school) life. Working toward a PhD is also a personal process through which you become an autonomous researcher with a distinct identity. This latter part of your training must be tailored to your goals and the needs of your future employer. Indeed, the success of your career depends on it.

This chapter explores career options for engineering doctors and suggests several strategies for kick-starting your reflection. You do not have to try them all. However, assuming that brainstorming about your future may wait is a mistake. Now is the right time to discover options that align with your interests and values. Your doctoral studies will span a long period: there will be plenty of opportunities for extracurricular activities. However, considering career options early might allow you to recognize

5.1 A few decades ago, when asked about their future careers, most PhD candidates answered *professor*. Nowadays, many consider other options from the get-go. Some have no idea; this chapter aims to broaden your horizons.

those you wish to seize and tailor your learning opportunities toward your future career.

Box 5.1

What to Do with a PhD?

The question I ask my students the most is: "What do you want to do with your doctorate?" I want to know because, as a faculty member, I come across several opportunities for my pupils. Knowing what each one is interested in allows me to pass one on to the student who is the most likely to accept it and benefit from it. I ask the question often because, quite frankly, I forget the answer, but also to allow the response to change over time. It is also possible that you have no idea what you want to do with a PhD, which is acceptable too. If this is your case, I suggest you use this chapter to brainstorm possible avenues.

5.1   WHAT TO DO WITH A PHD?

The list of possible career options for science and engineering doctors keeps creeping up. Here are common (and not-so-common) options:

university professor: tenure-track faculty member (from assistant to associate to full professor) responsible for teaching, researching, and performing community services (see figure 5.1);

lecturer and teaching-faculty: a faculty member who is involved in teaching and other services but does not typically do research;

researcher: an individual performing research in a university (as a researcher or as a non–tenure track professor), an industrial facility, or a governmental laboratory;

application specialist: industry liaison between high-tech solutions and end users;

project manager: individual responsible for the planning, procurement, and execution phases of a project;

university staff: from advising students applying for scholarships to organizing doctoral workshops, working in the research office, and administrating it, many university members hold a PhD;

consultant: someone who provides advice on their area of specialty, often within a team of experts assembled to tackle complex problems;

entrepreneur: a risk-taker who creates or operates businesses, sometimes leveraging from IP developed while in school;

science communicator: someone who popularizes science through conventional or new media. Examples include Dr. Neil deGrasse Tyson,[2] Prof. Brian Cox,[3] Prof. Brian Greene,[4] or Prof. David Suzuki;[5]

lobbyist: science advocates at funding agencies and government institutions who petition for better funding for R&D and influence policy decisions;

technology transfer officer: a person facilitating the translation of inventions toward innovation through the IP route. Some PhDs also transition to become patent agents;

head of state: the former chancellor of Germany Dr. Angela Merkel holds a PhD in quantum chemistry. She worked as a research scientist until 1989 when she entered politics;

astronaut: Dr. David Saint-Jacques[6]—a Canadian astronaut with the Canadian Space Agency (CSA)—received a PhD in astrophysics after completing a degree in engineering physics; and

superstar: Queen's Dr. Brian May holds a PhD in astrophysics, Prof. David Saltzberg is a science consultant for the television (TV) situation-comedy *The Big Bang Theory*, and, before winning the 2022 Nobel Prize in Chemistry, Dr. Carolyn Bertozzi played in a rock band with Rage Against the Machine guitarist Tom Morello.[7]

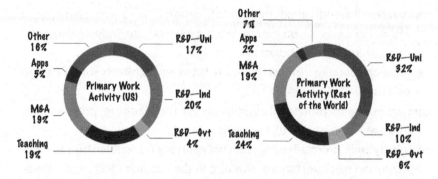

5.2 Typical careers for PhDs in science, engineering, and health fields working in the US (left) and abroad (right). Uni: university, Ind: industry, Gvt: government, M&A: management and administration, Apps: computer applications, Other: includes accounting, human resources, production, sales, quality management, professional services (health care, counseling, financial and legal services), and other activities not otherwise classified such as rock stars and astronauts.[8]

NOT EVERYONE IS A ROCK STAR

While some PhDs in STEM do turn into rock stars,[9] most choose more traditional options. Through regular surveys, the NSF is following a cohort of more than a million PhDs in science, engineering, and health fields who obtained their doctorate in the US. The 2019 edition shows that the odds of finding a job with a PhD are good. Indeed, less than 1.4 percent of respondents declared being employed, below the general unemployment figure for the same period. However, good jobs are odd. Yes, some of you will become university professors, but, according to *Nature Biotechnology*, for every ten PhDs awarded in STEM fields, only one faculty position opens up.[10] Most of you will therefore go on to another career. Figure 5.2 shows the distribution of STEM PhDs according to their primary activity, in the US (left) and abroad (right). While an important fraction continues as researchers in academic institutions, many pursue research in industry and government laboratories. Some even move on to management and administrative positions. This shows that—perhaps unbeknownst to you—PhDs have the skills to transition to positions outside academia.

When crossing paths with people occupying your dream job, scientist and author Peter S. Fiske suggests asking them three questions:[11]

- What made you choose such a career path?
- What steps did you take to transition into your current job?
- What classes should you have taken to prepare your transition better?

Box 5.2
Walk the Show Floor

Many engineering conferences have technical sessions where papers are presented and discussed as well as trade shows where vendors introduce their latest scientific instruments. In the market of highly specialized equipment, vendors often hold a PhD in engineering. In addition to discussing the specs of their latest products, you may want to ask about transitioning from academia to industry. Keep their business cards. If their company piques your curiosity at a later point, you may pick up the discussion with precise questions or tell a recruiter you already know someone inside the company.

5.2 ALL SORTS OF SKILLS

A doctorate makes you gain knowledge and acquire competencies, defined as how knowledge is exploited. Throughout your engineering studies, you are developing competencies belonging to two categories.

Core competencies are technical skills that define you as an engineer, such as translating problem statements into equations and concepts, solving mathematical equations, programming long pieces of code, assembling experimental equipment, designing experiments, making systems work, and so on. They constitute your strategic advantage in a technical environment.

Transversal and transferable competencies are essential to engineering and developed in parallel with core competencies. Communication, self-motivation, and discipline are examples of transversal skills. Skills developed in one context (academic research, for instance) and are helpful in another (industrial project management, for instance) are called transferable. Creativity, resilience, problem-solving, and complex analysis are examples of transferable skills from academia to industry.

Box 5.3

Transversal and Transferable Skills

Can you identify some transversal skills that you developed during your studies? Which ones do you think are important for a career in academia? Which skills could be transferred to a career outside academia?

ACADEMIA

Your PhD project and dissertation contribute to teaching you experimental and technical writing skills, respectively. A position in academia requires that you also show leadership, communication, mentoring, and management skills, to name a few. Indeed, university professors are more than researchers. To achieve tenure, they must demonstrate excellence in three areas:

Research: They must formulate original ideas to win competitive grants, support a team of junior researchers, and acquire and maintain instruments required to produce scientific results and advance the frontier of knowledge. As part of a broader research community, they review publications and grants from their peers (and competitors!). Professors are also required to initiate IP protection and assist in transferring technology to the next player in the chain of innovation.

Teaching: They train the next generation of engineers and scientists through lectures and lab work. They must prepare class notes, slides, homework assignments, and exams and grade them. While professors are explicitly trained for research, few receive formal training for teaching. If a professorship interests you, try to read about pedagogical strategies[12] or seek teaching experience during your studies. Some universities even offer pedagogy classes to their doctoral students. Teaching is the core of the job and not just a chore one reluctantly executes in exchange for the privilege of doing research.

Service is a broad term encompassing administration, community services, and outreach. Indeed, universities are administered by their faculty members at every level, from managing research laboratories to heading departments and organizing education programs. Faculty members also offer their time as conference organizers and journal reviewers and editors. They participate in community outreach by communicating (thus popularizing) their research results and general scientific concepts to a broader audience.

The scope of a professorship is much broader than experimenting in the lab and, thus, requires an extensive repertoire of technical and transversal skills.

INDUSTRY

Even though you have spent most of your life in school, as a PhD candidate, there are transferable skills you already possess that are useful for industry:

accelerated learning: after spending nearly two decades in school, you have become an expert at learning, and learning fast;

logic and deduction: presented with a problem, you can find a solution, or at least ask the right questions to untangle the situation;

technical comprehension: you do not fear reading an operation manual or a technical document to figure out how devices work;

systems building: although the systems you build are often for single-use experiments, you are familiar with describing methodologies and protocols for others to reproduce;

data analysis: you know how to extract trends from large amounts of data;

elevated vocabulary: you learn new lingo fast and are not afraid of jargon;

record keeping: from lab book to versioning, you know how to keep track of your steps;

information seeking: you know how to get to the bottom of a question by exploring references and references' references. You are also unafraid of the unknown, seeing it as just another challenge to solve; and

internally motivated: PhD students are used to working alone with minimal guidance and find motivation from within: as long as their work is meaningful, they do not need trophies or recognition to work hard for long hours.[13]

In addition to these skills, I would add that some of you are very good at communicating, networking, teaching, mentoring, and handling multiple projects in parallel. However, most academics poorly understand

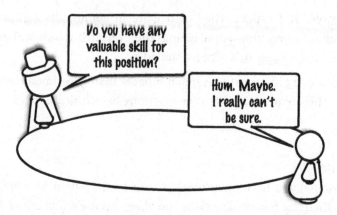

5.3 Identifying your aptitudes is a first step. Adjusting your attitude to the context is even better. Scientific humility is very important. But it needs to be contextualized: when presenting yourself to the outside world, be confident.

the industry (see figure 5.3): product and market knowledge, business acumen, and industry trends are not part of the usual repertoire.[14] Do you have some knowledge of products and markets? From the user's point of view, you do. For example, if you have used some instruments or some advanced software during your project and if you have assisted in purchasing by comparing them with competing products, you possess a glimpse of the market. My point is that you could, if you wanted, use your PhD to acquire some understanding of the industry, as long as you identified such a skill as something you wish to improve.

Box 5.4

Death of a Salesperson's Social Life

Technical people selling highly specialized instruments often hold a PhD in STEM and travel the globe, meeting laboratory directors, staff, and students. Unfortunately, they seldom return home at night. While it is a great way to see the world, it is a sure way to kill one's social life. Here is where you could step in. You are local; you know where the cool spots are for a coffee, a beer, an impromptu open-mike jazz night, and so on. Be a good host; you and your guest will benefit from bilateral information exchange from one scientist to another. Indeed, traveling PhDs know what it takes to work in the industry and where the latest openings are. Some even have an expense account for that extra coffee or beer.

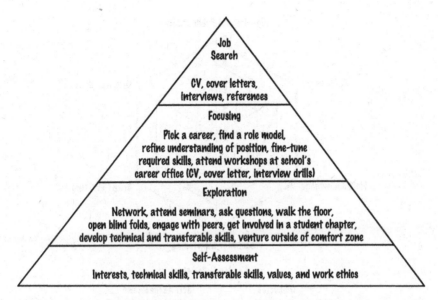

5.4 A match plan to prepare yourself for the after (school) life. Begin with a self-assessment based on the skills relevant to your future career. Explore, and engage with recent alumni from your group or your school working in various fields. Focus by projecting yourself on one of the possible choices and identify—when possible—a role model. Take advantage of the coaching from the career development office of your school, and go for it!

5.3 MATCH PLAN

Figure 5.4 describes a strategy developed by Peter S. Fiske in his book *Put Your Science to Work*[15] for preparing yourself for the after-PhD life. It begins with a self-assessment regarding your interests, skills, values, and work ethics. As you progress toward completing your doctorate, you will discover new interests and naturally develop new skills. But you could also deliberately seek opportunities to broaden your repertoire.

SELF-ASSESSMENT

Which transferable skills do you possess? Which ones are your weak points? Please take a moment to evaluate your strengths and weaknesses and compare them to what is expected of your future self. Figures 5.5 and 5.7 show two sets of skills, one for postulating to an academic position and the other for an industry position, as an example of a career outside academia. The edges of the radar plot show your objective, while the

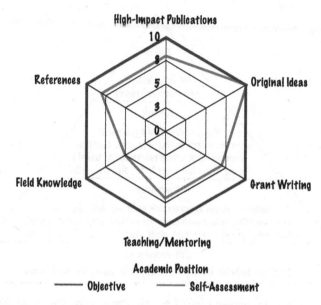

5.5 A skill set for academia. Outer radial plot: your objective. Inner radial plot: some hypothetical self-assessment. Individual skills may vary; the important point is to identify strategies for improvement.

inner line shows the result of a hypothetical self-assessment. For an academic appointment, having original ideas and high-impact publications is paramount. Still, other criteria are also considered: Do you have a good knowledge of your research field, have you trained or mentored pupils, have you participated in preparing a grant proposal, and do you have great reference letters?

Evaluating proficiency involves creating an evaluation grid. For example, an evaluation grid for scientific communications might look like:[16]

basic level: contributed to the writing of some sections of a manuscript; intermediate level: autonomously wrote all sections of a manuscript; or advanced level: edited the manuscript of another student.

For academic criteria, your advisor may help design milestones and evaluate you. You should consult counselors at your school's career development office for careers outside academia. The idea here is to identify possible shortcomings, develop intermediate steps, and devise strategies to improve your professional portfolio.

NO TRANSCRIPTS FOR SKILLS

To a certain extent, knowledge is assessed by your school transcript. There exists no equivalent for transferable skills. Your future employer (academic, industrial, or otherwise) must discover them via interviews and references. A bulleted list of skills is of little interest: anyone could fill a meaningless list. To be considered acquired, you must demonstrate each skill through achievements and behaviors and, ideally, reported by a referee.

Box 5.5
Asking for a *Great* Reference

Never ask for a reference letter, or you might end up with an ugly one (see figure 5.6). Instead, ask for a **great** reference letter. If the answer is no, move on to the next referee. If the answer is yes, make sure to provide enough context (the type of position), information about yourself [a curriculum vitae (CV) or résumé and a transcript], and your essay (if you are applying for a scholarship or another academic program). Add a few noteworthy points in the email accompanying your request, especially if your relationship spans many years. For example:

- I mentored an intern whose work was presented at CONFERENCE (YEAR).
- I got an A in your class NUMBER for my project on THEME.
- I was presented with the best poster award for my work TITLE at CONFERENCE (YEAR).
- My paper TITLE (JOURNAL, YEAR) is already cited NUMBER times.
- I showed leadership by organizing and running the journal club.

Bullet points are great; they allow your referee to wrap the facts with their writing style. Do not copy your CV, but curate a list of facts and skills you would like them to highlight. Splitting highlights among several referees is an excellent strategy to expose a complete portrait of yourself.

The radar plot for industry (shown in figure 5.7) is slightly different than for academia: industry contacts and knowledge play an essential role in finding a position outside of academia, and so do several skills such as good communication, data analysis, systems building, and being internally motivated. Your self-assessment for a position outside academia should be tailored to the employer's needs. Once more, your school's

5.6 Unconscious biases (see chapter 12) affect reference letters. Top left is an impersonal letter that was poorly copied and pasted from a previous one written for a different student without modifying fonts or pronouns. Such a letter is worse than no letter at all. A bad letter (top right) can fall prey to all sorts of common cognitive biases without providing evidence of merit or scholarship. For example, a review of 300 letters for medical doctors showed that letters written for female or male applicants differed significantly in length, features (accomplishments versus personality), and lexical fields (*he is a superstar* versus *she cares*[16]). A good letter (bottom) is personalized and emphasizes the uniqueness of the candidate's accomplishments.

career development office can assist you in identifying the ensemble of skills to begin your self-assessment.

INTERESTS AND VALUES
Career options for a PhD graduate are plenty, provided that you develop the right set of skills. You have the luxury of choosing a career that

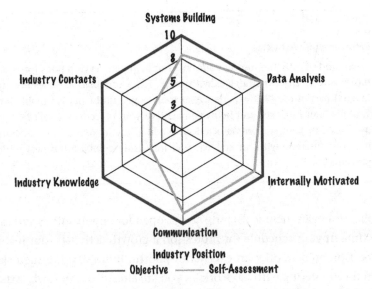

5.7 The ideal skill set (black) and hypothetical self-assessment (gray) for industry.

interests you in an environment respecting your values and matching your work ethic. For many of you, the problem is not saying yes to one interest; it is saying no to all others. In the past two decades, I have already met a few *once-in-a-lifetime* students: straight As, published in top-tiered journals while playing the first violin in the local orchestra between two Ironman triathlons. They are great at everything they do and capable of finding interest in every topic thrown at them, as long as a challenge exists. Yet, they have difficulty deciding on the One project to which to devote part of their careers. My prescription to them is to watch a different TED conference every morning until a theme strikes a cord. It does not have to be a TED talk, per se, but some rapid exposure to current research trends and challenges: a departmental seminar, a coffee with a colleague, a conference, or a networking event. Take a break from the lab once in a while to meet friends and network with future colleagues.

EXPLORE

Figure 5.4 suggests that after some initial self-assessment, you use your doctoral years to explore career paths and gain some of the skills you identified as your shortcomings.

Box 5.6

Extracurricular Activities

What kind of activities during your PhD can help you develop skills for your future career? Discuss strategies with peers. Extracurricular activities are an integral part of the PhD experience, as long as you find what is suitable for you, fits your interests, and helps you achieve your objectives. Don't be the deadwood of your student association to add a line in your CV: by matching your interests with an activity, you may truly combine business with pleasure.

Global strategies include keeping an eye open for opportunities and carving time in your schedule for professional growth activities, not just lab work. Conferences offer an excellent opportunity to develop such skills without interfering with lab work, as you are already out of town. Attend technical sessions during the day, but take advantage of the numerous evening activities offered to students. When none exist, take the initiative to schedule informal gatherings with other academic or industry leaders. Depending on the avenue you wish to explore, you may:

prepare for an academic position by asking your advisor to let you mentor interns, participate in teaching a class, prepare a literature review for a grant application, organize a conference, and review scholarship proposals of more junior lab members. Not all at once, and always in exchange for feedback (and perhaps money if you actually teach a class); or

prepare for an industry position by, first, not taking advice from academics who have never set foot in the industry! In his book, *Turning Science into Things People Need*, David M. Giltner proposes interviews of engineers and scientists who have left research laboratories to successfully pursue careers in the industry.[17] You may also try to get involved in a project with industrial partners, consider adding an internship to your studies, attend workshops from the corporate program of your usual conference, get involved in student chapters of your professional associations, and befriend traveling technical salespeople. Some faculty members who come from the industry may also help you navigate an eventual transition.

Box 5.7
Exploring for Introverts

Depending on your personality, you may wish to skip large gatherings and invest energy in one-on-one meetings. Conferences publish the list of the keynote, invited, and contributed speakers. Ahead of time, make a few appointments for coffee in a quiet environment, with a few targeted questions to break the ice ("I enjoyed this part of your presentation," "Your recent paper shows interesting data," and so on). One-on-one meetings are not just great for introverts but also for nonnative English speakers who benefit from the greater SNR of a quiet place.

FOCUS

After roaming outside of your comfort zone for the first half of your PhD, some of you will develop a mental image of your ideal position and will have met some inspiring leaders, possibly role models.

Box 5.8
On Role Models

Finding a role model is a double-edged sword: it is inspiring and intimidating. Meeting accomplished people may leave you feeling like you haven't done enough and will never be good enough. On the other hand, remember that the picture of a role model now is a maximum-intensity projection of the successes spanning their entire career: you hear of all their highlights—spread over several decades—within a one-hour keynote presentation. Please do not compare your career now to their early career then, but get inspiration from their work ethics, enthusiasm for the field, and core values. A typical academic CV spans several tens of pages. Yet, it only includes successful events. My CV would probably be twice as long if I included failed attempts at grants, publications, awards, positions, school programs, and so on. Indeed, Princeton's Prof. Johannes Haushofer (in)famously published his CV of failures, following advice from Dr. Melanie I. Stefan. Ironically, his résumé became viral and received *more attention than his entire body of academic work.*[18]

   Not finding a role model is also intimidating: it may make you feel like you do not belong. You do: it was not a mistake that you were admitted to your doctoral program, it will not be a mistake when you pass your comprehensive exam, and you can trust a jury of experts granting you a doctoral degree. If you are part of an underrepresented group in your community, you may seize the opportunity to act as the next generation's role model.

5.4 NETWORKING

Some PhD students in engineering are more comfortable designing the next generation of telecommunication network than actually using it for entertaining connections with other humans. For some, networking is akin to progressing through Dante's nine circles of inferno, which inspired figure 5.8. Each circle describes a new level of networking to conquer toward maximizing the likelihood of finding your X. From the center outward, the circles represent:

You: Why would others want to know you? Can you think of one serious, one fun, and one silly thing to say about you?

Your advisor: At first, your advisor is your boss; soon, they will become your colleague, and with a little luck, they will become a lifelong friend. Without being inquisitive, make sure your communications are at minimum cordial. There is a world between these two emails "sign attached document" and "Good morning Caroline, I hope you are well. Could you please sign this?"[19] Take advantage of a long drive to a conference to bond—conversing is often better than suffering silently through your advisor's outdated playlist.

Lab members: since you will spend more time with them than with your significant other, perhaps you should find some common interests

5.8 The nine circles of the networking inferno.

over lunch (see the invention of the global positioning system (GPS) below).

Departmental members: department seminars are a great way to enjoy free food and the company of other like-minded people.[20] Practice your networking skills by trying to make a short conversation with a different postdoc or faculty before each presentation starts.[21] At the end of a talk, ask a question, and stick around to chat with the speaker if the answer prompts a follow-up discussion.

Other schoolmates: chances are some of your classes include students from other departments—an excellent opportunity to practice describing your project in a lay person's voice.

Local chapter students: through local chapters of professional societies, you will meet students from other local institutions and, perhaps, through networking events, local professionals from industry and teaching institutions.

National community members: Funding of research projects occurs at the national level through government agencies and private foundations. Researchers from your country become reviewers of your funding application and collaborators on (or competitors for) grant applications. Part of this national community consists in industry members who benefit from a country's R&D and are asked to contribute financially to research funding. In the national circle, you will also meet policymakers looking for convincing arguments to improve governmental R&D funding, technology transfer office (TTO) employees trying to evaluate the potential of new technologies, as well as venture capitalists looking for the next unicorn.[22]

International community members: international conferences are where you can benchmark your latest results against the frontier of knowledge and meet leaders in the industry, as well as future colleagues from abroad.

Virtual community members: many academics are active on social media. Some brag about their glamorous lives, others about their recent publications.

Box 5.9

Global Positioning System

The GPS was invented by William Guier and George Weiffenbach, two lab mates at Johns Hopkins University's Applied Physics Laboratory (APL) who, after the launch of the first Sputnik satellite, decided to track its radio transmissions.[23] Guier and Weiffenbach wrote: "The Monday after the launch of Sputnik I, we met in the cafeteria for lunch." Before dinner, they had managed to capt the signal through a custom-built (hacked?) antenna and decided to analyze the Doppler shift for positioning. Soon, their preliminary data served as the basis for a research program resulting in the GPS. Dr. Emmett L. Brown and Sir Isaac Newton[24] aside, very few scientists experience breakthroughs when alone after hitting their heads. Instead, most discoveries happen from connecting dots when discussing with colleagues.

NETWORKING FOR ENGINEERS

Some of us are intimidated by networking. A common grievance I hear often is "I don't just know what to say." Here is a trade secret: you can network by saying almost nothing. Instead of staring at each other's shoes, or pretending you care about weather and gas prices, begin with a simple question: "What are you working on these days?" Indeed, most scientists are passionate about their work; that is, they spend so many hours in the lab to be able to entertain small talk on any other topic. Concerning the nine circles of networking: the smaller the circle, the closer your interlocutor's research is to yours, and the more likely you will be able to follow up with a relevant question and get the conversation going. If you cannot think of anything to add, try this line: "This looks hard!" You cannot be wrong: research is tough, and, as a bonus, you demonstrate empathy. Networking is kind, genuine, and generous: most people feel as vulnerable as you do. Inquire about someone's project, and you might learn something. If you hold the mike for too long, you only repeat what you already know. It is ok to ask simple questions[25] as long as you are genuinely trying to learn something. Networking is not about entertaining—you are not trying to bluff people with stuff you know—it is about creating connections.[26] Networking is also a two-way street: the more you give, the more you get in return. Letting a colleague know about a paper they might find interesting, a job opening in their field, or a speaking opportunity at a conference fosters professional ties.

Remark 5.1: In Case of Doubt, Smile *No one needs to know how uncomfortable you are. If you are neither interested nor engaged in the conversation, take advantage of the first occasion to move on to a different circle.*

Of course, you are not only networking with other PhD students. Often, you will meet with researchers from academia, industry, or government research labs. If you know ahead of time that you will be meeting a more senior scientist—for instance, if you are invited to lunch with the visiting speaker at the department's seminar—it is good practice to read one of their papers ahead of time. You may use the conversation to get an additional explanation about their project. If you don't know whom you will be meeting—for instance, when waiting in line for food at a large conference reception—it is totally fine to ask what is new and exciting in their lab. However, be prepared to answer the same question back: this ice-breaking question is a well-known networking trick. A simple "Why?" or "Why not?" adds fuel to the conversation. Pay attention not to be overly inquisitive, and offer some information about your project as well. Weather forecasts and sports results also work but result in much less memorable encounters.

Remark 5.2: Name Tag *No one is that famous: wear your name tag. More people than you know suffer from anomic aphasia. And if hand-shaking ever returns after the pandemic, wear your tag on the right side:*[27] *it improves visibility as you extend your right arm to greet someone.*

Finally, you may meet nontechnical people, for example, human resource (HR) specialists, in the context of a job interview. You will be asked to describe your project. Do prepare a lay person twenty-second version of the key research question and objectives you are chasing. You can rehearse it with family members during holiday gatherings. Your pitch is perfect when you avoid being asked next "When are you getting married?"[28] Every decade, the pendulum swings between including (or not) hobbies on your CV. One line mentioning activities outside the lab is a convenient ice breaker before the formal part of any interview.

THE HIDDEN JOB MARKET

The hidden job market describes positions that are not advertised but filled by word of mouth and targeted spontaneous candidacies. *Nature* reports that, in biotechnologies, up to 70 percent of entry-level industry positions are filled through word of mouth.[29] As a faculty, I witness a similar trend as I am often directly contacted by employers looking for candidates for their openings. As an entrepreneur, I fill positions both ways: roughly half our team was hired from our direct network, and the other half after a job had been widely advertised.

Remark 5.3: Faculty Positions *Contrary to some openings in the industry, faculty positions are always, always, advertised most of the time at an international level through trade journals and specialized websites.[30] However, since junior scientists around the globe do not all read the same journals, word of mouth also plays a role. Indeed, ads circulate informally from one faculty to their contact list and on social media. Some websites also scout recruitment ads and compile lists emailed to their subscribers.[31]*

5.5 THE NEXT STEP

TO POSTDOC OR NOT TO POSTDOC?

As you near the end of your doctoral degree, you may face every graduate student's dilemma: to postdoc or not to postdoc?

Remark 5.4: What Is a Postdoc? *Contrary to what its name might suggest, a postdoc is not a degree. It is a transient status between that of a student and an independent scientist.*

There are plenty of reasons for wanting a postdoctoral position:

the ugly, inertia: by the time you graduate, academia will start to feel very comfortable. Inertia alone might be playing a significant role in your wanting to seek a postdoctoral position;

the bad, survival: it might be the only immediate *job* offer you receive, and, while a postdoc does not pay much, it is enough to complement the free food from seminars with actual nutrients. Wanting to eat is not bad in and of itself, but with a PhD in engineering in your hands, you are entitled to entertain slightly greater ambitions;

the good, becoming a world-renowned academic scientist: inertia and survival aside, the good reason for wanting to postdoc is to become the best in the world at something unique. Are you postdoc-ing in the same lab to pursue an experiment that you dreamed of during your PhD AND your current lab is the best in the world to conduct such investigation? Are you, instead, bringing your newly acquired expertise to another lab to contribute to another subfield, and together, these two topics make you uniquely qualified to start your laboratory? This last reason is the best one for seeking a postdoctoral position.

Remark 5.5: The Two-Body Problem   *It is common to accept a temporary postdoctoral position in the same laboratory to finish an experiment or to wait for a partner to finish their degree before making a move together. Most employers are aware of the difficulties of finding two positions for highly qualified personnel in a given geographical area and, when asked, may lend a hand to find opportunities.*

To pursue a career in the industry, a postdoc is not necessarily an advantage in all STEM disciplines,[32] and you might want to apply to an industrial position directly. Unsure about the transition? Some countries fund industrial postdoc scholarships to allow straddling academia and industry for a few years before making the final plunge. Some fear that a postdoctoral position leaves you overqualified for a role outside of academia. It all depends on how you prepare your résumé (highlighting skills versus diplomas) and how you cultivate your transversal skill set and network during your fellowship.

Remark 5.6: Few Moves Are Ever Final   *I have met PhDs who have left academia to become entrepreneurs or even journalists and, later on, decided to complete postdoctoral fellowship before finally embarking on very successful academic careers. Inversely, some hard-core academic engineers and scientists discover a passion for entrepreneurship (a few even contributed to developing*

vaccines that saved the world from a global pandemic—thank you!). There are
no shortcuts to happiness, but there is also no point in staying in a position that
does not get you closer to your goals, that does not get you closer to happiness.

THE POSTDOC QUADRANGLE

If you do decide to pursue the postdoctoral route, you need to consider
the postdoc quadrangle:

the lab: find a place where you will learn a skill that makes your CV
   unique, under the guidance of a mentor whom you respect and who
   cares about the future career of their pupils;

location: your life outside the lab contributes to choosing your next
   location. You may find value in countries with family-friendly policies, want to get closer to aging family members, or have to juggle a
   two-body challenge;

money: postdoctoral positions open when grants are awarded and must
   be filled quickly to reach the grant's milestones within the prescribed
   timeline, restricting the hiring period to a tiny window; and

timeline: a postdoctoral position is supposed to be a transitional status;
   some funding agencies even cap the number of years you may spend in
   such a role. This does not mean that you shouldn't have rights during
   this in-between period: postdoctoral associations (and even unions)
   have successfully petitioned for minimum salaries and benefits: make
   sure you inquire about the whole package and do not sell yourself (too)
   short.

Finding a position matching all criteria involves some luck. However, you
may improve the odds by applying for postdoctoral scholarships while
developing an extensive network of future mentors to work in cities that
accommodate your ambitious goals and your sparkling personal life.

LIFE OUTSIDE ACADEMIA

It was common for the generation before *ours* to work for the same
company throughout their career. It is now quite unlikely that you will
undergo only one job interview in your lifetime. However, only while
in school can you benefit from the free help of top career counselors

specializing in engineering higher education. Use their assistance on several topics ranging from writing an effective cover letter and preparing a CV or résumé to practicing for a formal interview. Transitioning from the academic CV (Latin for *course of life*) listing diplomas, publications, and achievements to the industrial résumé (French for summary) focusing on competencies and vision is difficult: do not underestimate the task's importance and do take advantage of your school's resources. The résumé is indeed much shorter, and chances are that you will not be able to fit in (nor should you) the details of your numerous publications. This is where social media plays a (positive) role. Maintaining accounts on Google Scholar (for publications) and LinkedIn (for everything else professional) is an easy way to point an interested reader toward a comprehensive list of your articles and work without the burden of a thirty-page CV. What are the sections of a résumé? Ask your school's career counselor, as the answer varies with country and job type.

Box 5.10
Profile Picture and Online Presence

In the spirit of explicitly mentioning implicit notions, I will say that the half picture of you and your ex at a fraternity party does not quite convey professionalism. Perhaps your school or professional society student chapter organizes the visit of a professional photographer to help upgrade your profile at a low cost. Alternatively, ask a friend to take a well-lit picture of you against a neutral background. Use it for your professional social media accounts, but not on résumés in North America. Keep your online presence as clean as possible, including your personal email address: avoid using the likes of partyanimal@goodtimes.com.

Properly applying for a position is a lot of work and may generate much stress. As such, you should gather as much information about the position—through websites and your network—to try to gauge whether it is a good fit for you and, equally importantly, whether you are a good fit for the company. You may also try to learn about the company culture, policies, and values and assess if they align with your idea of a stimulating work environment. Inquire about salary, benefits, hours, vacations,[33] and parental leave. Get a sense of the opportunities for training and growth.

Is there a set path in the company for people with your skills, or is that not really part of their *modus operandi*? What is the company's policy for publications, conferences, and patents? Again, your school's career office may guide you in drawing an accurate portrayal of the position.

As previously said, most industrial highly qualified personnel (HQP) entry-level positions are not advertised but filled through word of mouth. This emphasizes the necessity of building a strong network on either side of the force. Your advisor might hear of industrial (or national laboratory or other) positions, and your school might host a career day inviting industries to meet with their graduates. Opportunities outside of academia are numerous. Use your PhD to explore as many facets of your future career as possible and build a network that will last a lifetime.

II

LEADING A RESEARCH PROJECT

6

MANAGEMENT OF A RESEARCH PROJECT

"If we knew what it is we were doing, it would not be called research. Would it?"

Albert Einstein

This book opened up with the high attrition rates observed worldwide as being caused by a combination of factors, including a poor grasp of the nature of the PhD and a lack of project management skills. Part I cleared common misconceptions regarding doctoral studies. Either you decided that it was not for you, or, indeed, you enrolled in a doctoral program. From this point on, let us assume you are on board and want to understand better how to structure and conduct your project.

Part II introduces methods to lead a doctoral research project in engineering. This first chapter describes (and, perhaps, justifies) project management in the context of a doctorate degree in engineering. It concludes with the four stages of a research project, each further explored in one of the following chapters. You will acquire the tools to refine, plan, execute, and conclude your thesis. Indeed, the authors of the book *The Management of a Student Research Project*[1] remark that "[students should not use] the inevitability of uncertainty as an excuse for not adopting a systematic and logical approach to their work."

Each project management step will also serve as an excuse to introduce topics relevant to doctoral studies in engineering: impact of research, sustainable development, time and priority management, risk analysis, ethics, and intellectual property protection.

6.1 A PROJECT

What is a project? The Project Management Institute[2] suggests the following definition: "a temporary endeavor undertaken to create a unique product or service."

This definition highlights the finite temporal dimension of a project: it has a beginning and an end (and sometimes, an extension to the planned ending that some, inaccurately so, call a postdoc). For a doctoral project, the beginning is known, but the end is not: it is, at best, estimated.[3] A doctoral project ends when a contribution to knowledge is made, not after having served an X-year sentence (despite the ending being decided on by a jury). Some universities have stricter rules regarding how many years may be devoted to doctoral studies, but reaching the end of these years does not guarantee a doctorate.

A project also aims to create a unique output, which, in the context of a PhD, is the doctoral dissertation. The notion of producing a unique output comes in opposition to plowing through mundane activities. In their reference work *Project Management Handbook*, Cleland and King further add the notion of complexity to the definition of a project.

Definition 6.1: Project *A project is a **complex** endeavor to achieve a unique and somewhat ambitious goal within a time frame and budget. The output of a project is a unique product, process, or result. A project has a beginning and an end: it must be distinguished from routine operations. A project is not an ensemble of mundane tasks, nor is it indeterminate in time and required resource.*[4]

In other words, going to class, preparing homework, reviewing the scientific literature, and performing experiments for someone else do not constitute a project, let alone a doctoral project. Such tasks are not uniquely tailored to a particular goal. A project is never a collection of

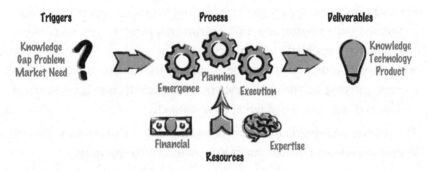

Triggers   **Process**   **Deliverables**

Knowledge Gap Problem Market Need

Planning
Emergence   Execution

Knowledge Technology Product

Financial   Expertise
Resources

6.1 From left to right: a research project is prompted by a trigger—some knowledge gap, problem, or market need—and processes resources—time, money, and expertise—into deliverables—knowledge, technology, or products.

unrelated prosaic tasks. Rather, it is a complex ensemble of structured activities planned and realized to achieve a specific mandate leading to an anticipated deliverable within a prescribed, albeit flexible, time frame. Figure 6.1 illustrates the project from its triggering points to transforming resources into deliverables.

A RESEARCH PROJECT IN ENGINEERING

Popular culture has portrayed a relatively slow-paced, romantic view of the academic researcher hanging out in coffee shops (Hello Dr. Ross Geller![5]) in search of inspiration. The reality of engineering graduate students is quite different, except for the not-so-comical large doses of caffeine often involved. A PhD requires completing a research project to advance knowledge, develop technology, and, sometimes, promote innovation. As a project, doctoral research:

has a beginning and an end: do not wait for classes to be over before getting started in the lab, and beware of time limits imposed by your school, funding sources, and visa;

proposes unique deliverables: the dissertation is one of the requirements, and, in addition, many doctoral students author scientific publications and even patents;

is complex: expect to be working on a non-trivial topic and on a question to which, by definition, no one knows the answer;

requires resources: the project requires the time and expertise of yourself and your advisor, and, for experimental projects, you also require equipment, materials, and money; and

is not a mashup of routine operations: simply showing up to group meetings, carrying on literature reviews, or, especially, watching YouTube videos is not considered doing your research.

The triggers and deliverables vary according to the nature of the project within the research-development-innovation (RDI) spectrum.

6.2 THE RESEARCH-DEVELOPMENT-INNOVATION SPECTRUM

In engineering, pushing the boundary of knowledge is performed in several ways. Some students tackle fundamental science questions, others develop new instruments, and some are involved with introducing novel ways to promote innovation. In chapter 7, we will review models of research in a way to elucidate where novelty may be found. Here, we describe a simple, linear continuum encompassing research, development, and innovation. The research project of a PhD candidate in engineering indeed stands somewhere within the RDI spectrum.

Box 6.1

The Laser Pointer

Figure 6.2 shows the evolution of the laser along the RDI spectrum. At the far left, research in quantum mechanics revealed the energy structure of some atoms. From this new knowledge, engineers obtained stimulated emission of radiation, a laser, in a laboratory environment. Further research established that compact semiconductors could also produce laser light. Several rounds of developments later, the compact laser pointer was born. Thanks to this massive effort in RDI, presenters now have a useful tool to show data (and entertain pets).

TECHNOLOGY READINESS LEVELS

Figure 6.3 introduces the technology readiness level (TRL) scale. Developed by the National Aeronautics and Space Administration (NASA), the

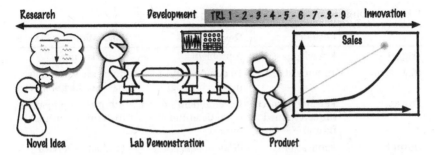

6.2 A laser pointer at every stage of the research-development-innovation spectrum. Left: a new theory related to quantum mechanics reveals quantized energy levels for atoms in gas. Center: engineering physics development allows laser light to be generated from the gas mixture (and later on from semiconductors). Right: innovators packaged the new instrument, the laser pointer, to the benefit of some end users. Not shown: a YouTuber is (ab)using the device to play with a cat.

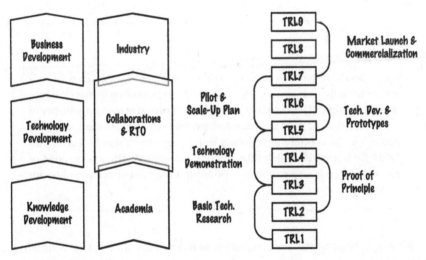

6.3 Technology readiness levels from (academic) knowledge development at level 1 to (industrial) business development at level 9.

TRL scale assesses the maturity of a new technology.[6] The nine-level scale ranges from early reports of new concepts (level 1) to proof of successful deployment of the technology in an operational setting (level 9). In other words, it ranges from the back-of-the-envelope calculation to innovation reaching end users.

Table 6.1   Dimensions of research projects in engineering[a]

	Research	Development	Innovation
Trigger	Knowledge gap	A problem	A market need
Objective	To understand	To demonstrate a new technology	To create value via a new product or service
Process	Research project (undetermined future)	Development project (semi-determined future)	Innovation project (determined future)
Output	Publication(s)	Proof of concept	Product or service
Stakeholders	Academia & governmental laboratories	Industry, academia, or governmental laboratories	Industry and RTOs
Potential flaw	Lack of originality	Poor design	Poorly targeted market

[a] Inspired from Yves Langhame, "CAP7015E Leading a Research Project" (class notes, 2015).

Box 6.2
TRL Scale

Where does your research project fit within the TRL scale? Discuss with your advisor to verify your answer. Does it matter? The level often dictates the type of funding and collaborations you may have. It may also indicate whether or not your project is a good candidate for an entrepreneurship venture after your studies. On the contrary, at very early levels, your project may be a good candidate for a scientific publication in a top-tier journal. There is no good or bad answer: each step within the TRL scale brings about opportunities for researchers.

Table 6.1 highlights particularities RDI projects along the following dimensions: triggers, objectives, process, deliverables, key players, and potential shortcomings. Such extremes are discussed for illustration purposes as the boundaries between research, development, and innovation are often blurred.

SCIENTIFIC RESEARCH

Triggered by a gap in knowledge, the scientific research project aims at creating a new understanding disclosed through peer-reviewed scientific publications. This research project is carried out in universities and government research centers, and its future is somewhat undetermined. Originality primes over immediate usefulness, although the two are not mutually exclusive. Potential shortcomings include results of unconvincing originality or significance or getting scooped by another team.

TECHNOLOGY DEVELOPMENT

Triggered by a potential application, technology development aims at demonstrating new technology, also disclosed through scientific publications, perhaps in parallel with invention disclosures and IP protection (more on this in chapter 10). The development project is carried out in universities, government research centers, and industries. Its future is semi-determined as new avenues may be discovered in the process. A notorious example is the fortuitous invention of microwave ovens by a radar scientist who noticed his chocolate bar melting as he turned the device on. The outcome of such a project is a proof of concept that could be published, patented, or both, and a potential shortcoming may be a nonfunctional or poorly designed prototype.

INNOVATION

Defined as the practical realization of a novel idea that results in a new product or service, innovation differs from development in that it reaches the end user. It is triggered by a market need and leads to the commercialization of new technology. Innovation is not concerned with furthering the understanding of fundamental questions or developing an intermediary instrument but rather aims at successfully materializing its applications.

Figure 6.4 illustrates four types of innovative work categorized into new and existing markets versus new and existing technologies:

architectural: when existing technologies are reorganized to open a new market (e.g., the original smartphone);

incremental: when an existing technology (e.g., an articulated head) is transposed to an existing market (e.g., facial hair management), and yet completely transforms behaviors (e.g., the way people shave);

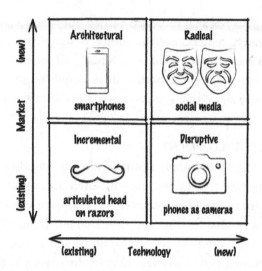

6.4 Innovation quadrants. Innovation may target either new or existing markets, using new or existing technologies, through architectural, incremental, radical, or disruptive qualities.

disruptive: when a new technology (e.g., compact lenses within new smartphones) shake up an existing market (e.g., photography); and

radical: when a new technology opens up a new market (e.g., AI enabling social media) (Yes, there existed a *before* social media).

Innovation can be the motor of a doctoral project as long as there exists a theoretical framework supporting the project and leading to new knowledge. Examples include: micro-electromechanical systems (MEMS)-based gyroscopes and miniaturized telephoto lenses within smartphones, face recognition or other novel AI algorithms behind social media, and so on. However, designing the contours of the new smartphone or razor is not a doctoral project in engineering, unless, perhaps, it is associated with a fundamental research question linking design parameters with consumer behaviors.

While innovation may be found in universities, it mostly occurs in industrial settings, and within research and technology organization (RTO)s. Its future is determined, and one of its shortcomings is a poorly targeted market analysis.

Definition 6.2: Research and Technology Organization *A nonprofit organization closely cooperating with industries, large and small, and several public actors to perform basic and applied research promoting innovation.*

Box 6.3
Your Project within the RDI Spectrum

Where does your project fall within the RDI spectrum? Identify your project's trigger, objective, process, deliverable, key players, and potential shortcomings. In what regard are you pushing at the boundary of knowledge?

Since doctoral projects must advance the frontier of knowledge, most PhD students' projects will revolve around research and development. However, some engineering fields are concerned with the process and management of innovation. Thus, in this chapter and the next, we will describe all three dimensions of the RDI spectrum. As a student, however, it is your responsibility to seek avenues leading to new knowledge. Bringing a product to market is a tremendous exploit, but it does not generally constitute a doctoral contribution.

6.3 PROJECT MANAGEMENT

Given the uncertainties in science, is project management desirable or even feasible? Project management is intended to be flexible and accommodating. It encourages a systematic review of the objectives and the situation and a continuous quest for the best path forward. The management of research projects involves exploiting all available resources to optimize the outcome. While project management might be seen as yet another skill to master, it is first and foremost a culture concerning the definition, planning, conducting, and conclusion of your research project. The few techniques described in the following chapters should suffice to attain noticeable results. The key is to adopt an attitude of openness toward learning skills other than purely scientific or technical ones.

Remark 6.1: Failure to Launch *Some students enjoy the management process so much that they risk paralysis through analysis. Remember, balance is*

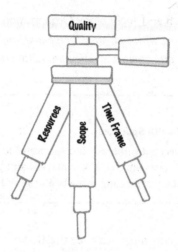

6.5 Inspired by the Golden triangle, the project management tripod is based on three pillars: scope, resources, and time frame.[9] Given a fixed leg, adjustments must be made to the other two to maintain quality.

paramount: no engineering degree is awarded for producing the most beautiful Gantt charts alone. As a rule of thumb, no more than 10 percent of a project should be allocated to its management, including group meetings.[7] Nevertheless, there is also a learning curve associated with project management. Therefore, you should not judge yourself too harshly if you were to bust the 10 percent limit in the early months of your project.

As suggested in *Making the Right Moves*, a management guide for scientists prepared by Howard Hughes Medical Institute (HHMI), project management is "a series of flexible and iterative steps through which you identify where you want to go and a reasonable way to get there."[8]

PROJECT MANAGEMENT TRIPOD

Managing a project is like balancing a tripod, where each leg represents the project's scope, resources, and time frame as illustrated in figure 6.5. Management begins by setting the length of the first leg. For example, one could work with a fixed deadline (e.g., the *time frame* leg). In order to complete the project (or achieve equilibrium for the tripod), the other two legs must be adjusted. In the current example, these two remaining legs are *resources* and *scope*. To meet a fixed deadline, one may either reduce the scope of the project or increase resources assigned to its realization. The same exercise holds true if another leg is prioritized. A student may, for example, choose that the scope is nonnegotiable. In this case, the student must agree to a longer PhD or increase the resources by delegating part of its data gathering to an intern. If hiring an intern is impossible, a compromise must be found between the perfect (scope) and finished (time frame) thesis.

Box 6.4
Golden Triangle

When trying to decide on the scope of your project, ask yourself the following questions. What is your ideal timeline for graduation? Would you be willing to mentor an intern to assist you in your project? What is most important to you: an ambitious scope, early graduation, working alone, or managing a team?

TRANSFERABLE SKILL

Several studies identify skills associated with project management as key expertise for researchers, even junior ones. The skills involved in research project management are transferable to your future career and include being able to:[10]

- formulate innovative research questions;
- perform multiply-disciplinary work;
- leverage from existing research, technologies, and knowledge;
- communicate as a way to motivate and mobilize a team;
- assess the progress made by the team;
- direct the efforts; and
- manage resources and partnerships.

6.4 STAGES OF A RESEARCH PROJECT

Figure 6.6 shows the four stages of a research project:

emergence and definition: discussed in chapter 7, the emergence phase is an iterative process of growing and pruning ideas, leading to the first draft of a research question, subsequently defined through research objectives, methods, and anticipated results;

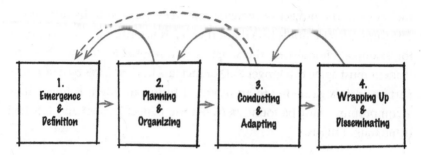

6.6 The four stages of a research project. The arrows illustrate the nonlinearity of the process: it is common to revisit the previous stage during the doctoral journey. The dashed arrow shows a not-so-uncommon case of starting over when an insurmountable hurdle is met.

planning and organizing: described in chapter 8, the planning phase con-
    sists in detailing all dimensions of the project and breaking these
    dimensions into work packages to which resources may be assigned;
conducting and adapting: in the conducting phase, research finally hap-
    pens starting with the kick-off meeting and evolving through regular
    progress assessment, risk analysis, and adaptation, as hurdles, new
    opportunities, and ethical questions inevitably arise; and
concluding and defending: in this final phase, the topic of chapter 10, the
    project is finalized, the thesis submitted, results protected for IP and
    disseminated, and know-how is transferred to the next student. As a
    side-effect, you get to graduate with a PhD.

The arrows in figure 6.6 illustrate the constant revisions of the research
project as new avenues emerge, and some sometimes reveal dead ends
from which researchers must pivot.

7

EMERGENCE AND DEFINITION

"A scientist describes what is; an engineer creates what never was."

Theodore von Kármán

Tucked between a trigger and the planning of a first experiment, the crucial emergence phase defines the most critical aspects of your quest for new knowledge. A thoughtful study design increases the likelihood of producing interesting results.

It also maximizes the odds of obtaining research funding. Through the emergence phase, the question is refined: Is it new, important, and timely? A hypothesis is posited: Is it plausible, measurable, and original? General and specific objectives are crafted: Is the project's mission aligned with the lab members' competencies? The scope is estimated: Where does the project begin and end? Finally, key assumptions are established, for example: How many terms in the Taylor expansion will be considered? In other words, how far do you need to go to model a cow: Is a sphere enough, or should you also add legs, a tail, and a head?

Important results typically stem from important questions. History has, of course, witnessed occasional serendipitous discoveries: from Petri dishes left unattended and leading to penicillin to large batches of not-so-strong glue turned into sticky notes.[1] However, relying on chance should not be the sole strategy in a researcher's playbook as it has seldom contributed to securing competitive funding. This chapter looks into a systematic approach to defining a research project in engineering. The goal is twofold: increasing the odds of your project resulting in novel and significant results and securing a large dose of motivation to keep you going for the long run. We will also look into serendipitous discoveries on the off chance that we could learn from them.

7.1 GENESIS OF A RESEARCH PROJECT IN ENGINEERING

As explained by David V. Thiel in his book *Research Methods for Engineers*, engineering research concerns applying the scientific method toward proposing realistic systems to the benefit of humankind.[2] The scientific method is an empirical process applied to knowledge generation combining induction and deduction:[3]

induction: careful observation leads to the emergence of patterns that form the basis for a hypothesis and, eventually, a theory. For example, a series of observations (e.g., *I suffer from anaphylactic shock when I eat shrimps*), followed by the emergence of a pattern (e.g., *anaphylaxis is a symptom of shrimp allergy*) leads to a hypothesis (e.g., *I may be allergic to shrimps*); and

deduction: intelligent consideration of a theory leads to a hypothesis either confirmed or informed by observation. For example, start with a theory (e.g., *women are mortal*), followed by an observation (e.g., *Caroline is a woman*), and leading to a conclusion (e.g., *Caroline is mortal*—sigh!).

Figure 7.1 shows a linear representation of the steps of the scientific method. The arrows illustrate that the process is hardly linear—several loops between individual steps are the norm rather than the exception. At the top, the quest for new knowledge begins with a trigger—some observation, a question, a starting idea—prompting the initial research (literature search, computations, and simulations coupled with a healthy dose of skepticism). From this initial investigation, a pattern emerges; a hypothesis or thesis statement regrouping question, hypothesis, general and specific objectives, assumptions, and scope. Experiments are planned

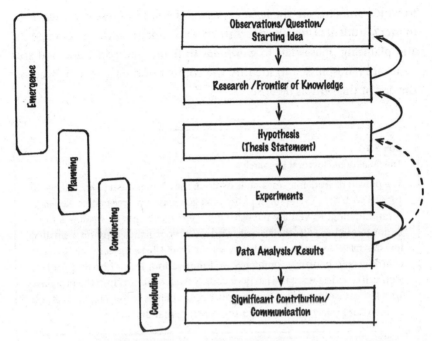

7.1 Following the scientific method, the PhD process involves turning an idea, the starting point, into an original and significant contribution.

and conducted, and data are analyzed with as many iterations as are needed to achieve robust and reproducible results from which a conclusion is reached. Communicating these results to a community of peers allows for including the findings in a verified body of literature for the next research cycle.

THE SCIENTIFIC METHOD

In figure 7.1, the scientific method is juxtaposed to the four phases of a research project. The emergence phase spans the first three steps of the scientific method and largely establishes the value of the project: Is the question important? Is the hypothesis weak or hardly plausible? Are the objectives aligned with the core skills of the research team? Is the need pertinent to the expectations of industrial partners? No amount of planning or careful execution can compensate for a poorly designed study. Emergence is the most crucial stage of a research project, yet, it is the stage to which students are the least exposed. The tasks of research interns or master's students often begin with an experiment, with the emergence and planning phases left to *grown-ups* to figure out. Some doctoral students are hard-pressed to identify the context of their research, even on the day of their thesis defense.

Box 7.1

The Duality of Doctoral Research

The pursuit of academic research is twofold: the development of knowledge in parallel with the development of the next generation of researchers. Research is intimidating. As you are navigating the emergence phase, shadow a more senior researcher with their experiments. Get your hands dirty on a smaller, low-risk project to get a tangible feel for the field. Chase the early success that contributes to boosting confidence and motivation. Do not be that student who still wonders what the lab door code is twelve months into the program. Very few courses teach you how to splice optical fibers or use a pipette. As you define your project, also start figuring experimental things out.

In an ideal world, as the purpose of a PhD is also about training students to become autonomous researchers, doctoral candidates should take part in every step of the project, including its emergence. "...in theory there is no difference between theory and practice, while in practice there is."[4]

Engineering research, which is often equipment intensive, is funded through competitive grants involving an effort from the PI that is often inversely proportional to the success rate of the grant competition.

THE TIP OF THE ICEBERG

Figure 7.2 shows the full emergence phase of a research project in engineering. A starting idea is investigated with some limited funding (start-up funds for new faculty, leftovers from a previous grant, or some rare discretionary funds) to demonstrate a proof of concept and obtain preliminary data. From these, a grant is drafted and circulated

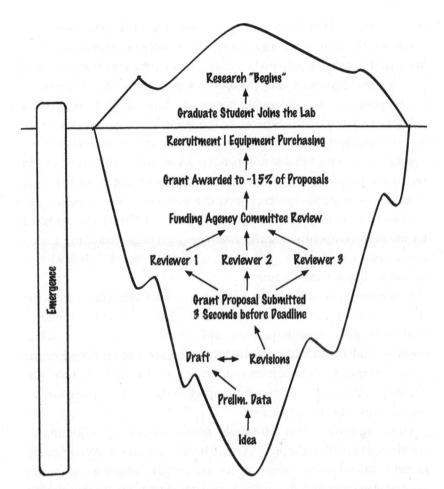

7.2 Emergence and the tip of the iceberg. While doctoral students benefit from participating in all project stages, they often join the team after the twelve to eighteen months required to obtain funding. Missing from the graphs are steps related to identifying the proper granting opportunity or specific to collaborative work (e.g., herding the team) or industrial partnerships (e.g., securing their contributions prior to submitting a grand and pre-negotiating IP considerations).

to colleagues (victims?) who volunteer avenues for improvement. The research office then verifies that the topic falls within the program's scope and that the budget is prepared according to the program's rules (and that page margins rigorously obey the agency's guidelines). The PI fine-tunes the proposal and accompanying documents until about three seconds before the submission deadline. From this point on, the grant agency evaluates the eligibility of the proposal and PI with respect to criteria from the agency, then finds a suitable number of peers willing to volunteer time to review the proposal. After a few months, all reports are presented to an evaluation committee, which selects the worthiest research projects to fund (with a success rate ranging from 50 percent to less than 10 percent for the most competitive grants). Awardees then begin recruiting a team and shopping (*again*[5]) for quotes to purchase equipment. This is when a graduate student joins a team.

Given the circumstances, the emergence and definition phase from the student's point of view becomes about catching up rather than coming up with the context, research question, and objectives of the project. When possible, students should be invited to peruse the grant proposal or the research contract. An important task remains for the student: highlighting the novelty and strengthening impact of the work, as these are the requirements of a doctoral thesis.

Other students—either self-funded from a scholarship or joining an established lab with an imposing research infrastructure—may participate in earlier phases of the emergence iceberg for full immersion. No matter at what stage you join the research project, the onus is on you to steer the project in the direction that maximizes originality and significance and the fit between the specific objectives and your core competencies (or competencies you wish to acquire). Discuss with your advisor to uncover niche opportunities.

7.2 SOURCES OF ORIGINALITY

One of the challenges of metamorphosing from user to creator of knowledge involves assimilating the scientific process of creation. With the proper resources and protagonists, the iterative algorithm of the scientific method may result in original and significant contributions (that must be communicated!). The student navigates these steps with the guidance of the advisor, thesis committee members, and collaborators, but the student, and the student alone, is responsible for mobilizing these resources. You are no longer a passenger but the pilot of your own doctoral journey.

What is an original contribution? One could argue that originality can be found at every step of the scientific method. Your starting point may be some new observation from which you identify a pattern allowing you to formulate an original question. Alternatively, you may also revisit an old question from a fresh perspective with a new hypothesis, using new tools and procedures or modern data analysis techniques.

STARTING IDEA

The starting idea for a research project may come from:

your thesis advisor: research theme, previous publications, previous student's theses, patents, grants, industrial contracts;

an external partner: an enterprise, a research center, a governmental agency, a call for proposal; or

you: the continuation of your master's research, a scientific question you have at heart, a topic to which you want to contribute.

For some, the idea is well mapped out from the get-go; for others, only a general direction is provided. The more detailed the project, the less academic freedom: choose a project that matches your tolerance to risk and uncertainty when possible.

Originality can be found in all components of the thesis statement.[6] You may indeed be:

- asking a question that has never been asked before. For example, after discovering the electron, Thomson inquired: How are charged particles organized within matter? He eventually proposed the plum pudding model in which electrons are scattered in a positively charged matrix;
- proposing a new hypothesis to an old question. For example, following Thomson, Rutherford also inquired about the atomic model but posited a different one: a mostly empty atom with densely packed positive nuclei and tiny electrons orbiting around it;

- exploring a new problem or area. For example, the idea of an atomic nucleus prompts the question: "What is it made of?" leading to the new field of nuclear engineering;
- suggesting a new theory. For example, Einstein's photoelectric experiment led to the theory of quantum mechanics;
- developing new objectives or applications never suspected before or suffered from technological limitations. For example, foundational experiments in quantum mechanics are revisited to produce—as opposed to studying—photons and, eventually, lasers;
- investigating a new methodology often inspired by the actual method or transposed from another field. For example, one may revisit the photoelectric effect with a different light source; or
- performing a new analysis leading to a new interpretation. For example, one could revisit old experimental data with new statistical methods.

Such examples were chosen as most engineers have some knowledge of modern physics, and not to convey an expectation that your PhD must result in a dramatic paradigm shift (and certainly not without a clutch![7]).

ORIGINALITY AT THE INTERSECTION OF DISCIPLINES

Some of the most original research occurs at the intersection between disciplines. In the broadest sense, research is split into disciplines: social sciences, humanities, life sciences, natural sciences, and engineering, and it is funded through distinct agencies or research councils, each funding research from one or two disciplines.[8] A fundamental problem with segregating research topics within silos is funding research relevant to many disciplines. For example, how do you fund biomedical engineering research, which solves life science problems using engineering tools? In some cases, research councils join forces and announce joint funding opportunities for research that is multidisciplinary, interdisciplinary, or transdisciplinary:[9]

Multidisciplinarity consists in solving a problem from different approaches, each within its boundaries. An example is the development of smart cars, which requires code from software engineers, sensors from

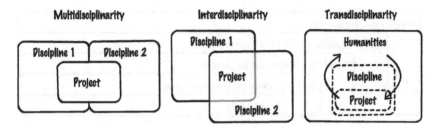

7.3 A graphical representation of multiply-disciplinary projects.

electrical engineers and engineering physicists, and novel guidelines from lawyers and philosophers. A multidisciplinary approach requires several experts, each contributing to distinct expertise.

Interdisciplinarity consists in solving a problem using knowledge from many disciplines onto a coordinated and coherent whole. MRNA vaccines against COVID are a fantastic example of interdisciplinary research involving work from life scientists as well as that of chemical engineers and physicists developing lipid nanoparticle delivery systems.[10,11] Interdisciplinary research leads to the training of experts with knowledge in both fields.

Transdisciplinarity is the integration of natural, social, and health sciences in a humanities context, with the transcendence of their traditional boundaries. Sustainable development is an example of transdisciplinary work as solutions from science, and engineering must receive social acceptance.[12] The collaboration between visual artist David Hockney and optics professor Charles Falco is another example of transdisciplinary work. Hockney wondered if some Renaissance masters had relied on instruments in their paintings. Prof. Falco, using his knowledge of optical aberrations from lenses and mirrors, calculated that the use of various vanishing points was consistent with the use of primitive cameras.[13] Here, an art history question was answered using tools of optics, and our knowledge of the use of technology was answered through visual arts.

Choi et al.[14] note, however, that these three terms all describe the interaction between different disciplines to varying degrees on the same continuum.

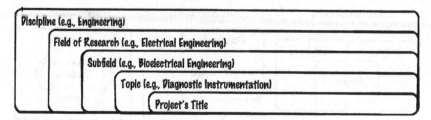

7.4 An example of progression from discipline, field, subfield, research topic, and, eventually, the project's title.

ORIGINALITY AT THE INTERSECTION OF FIELDS

Interesting problems are also found at the intersection of research fields, a subclassification within disciplines. Indeed, within a single discipline, the scope is still too broad. How can you compare the merits of projects in quantum communication and machine learning? To avoid them competing for the same pool of money, research councils define research fields.

FROM DISCIPLINE TO PROJECT'S TITLE

Figure 7.4 illustrates the progression from discipline to research topic, which eventually extends into the project's title. The fields, subfields, and topics are, to some extent, defined by funding agencies. For example, within science and engineering, Canada's Natural Science and Engineering Research Council (NSERC) defines research fields as:[15]

- Genes, Cells, and Molecules;
- Biological Systems and Functions;
- Evolution and Ecology;
- Chemistry;
- Physics;
- Geosciences;
- Computer Science;
- Mathematics and Statistics;
- Civil, Industrial, and Systems Engineering;
- **Electrical and Computer Engineering;**
- Materials and Chemical Engineering; and
- Mechanical Engineering.

Each research field is further divided into subfields. For example, the field *Electrical and Computer Engineering* encompasses:

- Computer Hardware and Architecture;
- **Bioelectrical Engineering;**
- Communications and Networks;
- Electromagnetics; Photonics and Micro/Nanotechnology;
- Electrical Energy Systems;
- Electrical Circuits and Electronics;
- Video and Display Technologies; and
- Signal Processing.

Granting agencies also provide examples of research topics within subfields. For instance, NSERC's website describes *Bioelectrical Engineering* as:

- **Biomedical and Diagnostic Instrumentation;**
- Medical Devices;
- Medical Robotics;
- Image-Guided Surgery Systems; and
- Biomedical Sensors.

Researchers rarely *choose* what to work on based on such categories; rather, they find the best fit for their research to be evaluated by the most competent colleagues. Indeed, funding proposals are—like publications—evaluated by a committee of peers, that is, other researchers in the same field. By selecting a field, researchers implicitly select which committee will evaluate their research application.

Box 7.2
Research Community

Research fields define communities of researchers with similar interests who gather to share ideas and resources[16] and peer-review grants and publications. Each field has one (or more) professional society that organizes meetings and publishes journals. For your field of research, identify which professional society is closest to your work, and perhaps, inquire about the possibility of joining as a student member or volunteer. Volunteering includes reviewing papers (starting as senior doctoral students), helping organize and host research conferences, and driving student chapters and outreach activities.

The progression detailed here is from my first grant on developing optical fibers for endoscopy. Despite being hired within the engineering physics department, I *chose* to be evaluated as an electrical engineer, as the committee was a better fit for my research topic. History has witnessed more striking, yet harmless, transformations, notably that of *physicist* Ernest Rutherford winning the 1908 Nobel Prize in *Chemistry*.[17]

Box 7.3
Emergence of Your Project—Part 1

Discuss the triggers of your research project and highlight the following:

- your discipline;
- your field of research;
- your subfield of research; and
- the topic of your research project.

Discuss with your advisor the sources of funding for your project. For example, which agency or industry funds your laboratory?
Advanced topic: Which professional society best serves your subfield of research?

When hesitating between two communities, discuss your options with your advisor and fellow students. Which community is the most supportive of its junior members? For example, which professional society provides scholarships, travel grants, and networking and professional development events?

Box 7.4
Take a Chance on the (Free) Cookie

In the mid-1990s, a technique called optical frequency domain reflectometry[18] was introduced to measure defects in fiber optics components. A few years later, it inspired a redesign of a medical imaging technique called optical coherence tomography (OCT).[19] Fast-forward to 2021, a new fiber optics device developed for OCT[20] finds applications in the LIDAR industry. If researchers gather according to their respective fields, how did innovation permeate such silos between telecommunications, medical optics, and ranging? Through departmental seminars. Academic departments span several fields within a specialty: a diversity represented in the variety of lecture topics. Take a chance on the seminar, even if it is not directly aligned with your research project. Worst-case scenario, you get a free cookie.

Finding originality in and of itself is relatively easy. The challenge is tackling a project that is both original, significant, and funded!

7.3 SOURCES OF IMPACT

An important success criterion for a doctoral dissertation is that it impacts knowledge and society. Students in my doctoral workshops are asked to complete these three sentences:

- the short-term impact of my project will be...
- the mid-term impact of my project will be...
- the long-term impact of my project will be...

Over the years, I came across a *few* wrong answers:

- leaving the sentence incomplete. If the answer does not come to you, consult your advisor. It is indeed important that together you identify some areas of influence;
- "the immediate impact of this project is me getting a PhD" is also wrong. The impact is measured as the effect your project has on furthering knowledge and on society, not just on yourself. Although I am the first to argue that getting a PhD is pretty cool; and
- allowing your lab, and only your lab, to progress in a particular direction. Be suspicious of projects interesting to no one outside your thesis advisor. Perhaps your advisor is a visionary ahead of contemporary engineers by a few centuries. Possibly. Most probably not. Trust Occam's razor and find a project for which you can satisfactorily complete all three sentences.

Some students argue that because their project falls at one point or another of the RDI spectrum, it might be impossible to identify the short– or, conversely, the long-term impact of their research. Updating the linear model of RDI alleviates such a common concern. For some projects, the immediate consequence is more readily identified than the long-term one, and vice versa.

> **Box 7.5**
> **Fusion Power**
>
> In December 2022, the US Department of Energy (DOE) announced the achievement of fusion ignition at Lawrence Livermore National Laboratory.[21] While the governing principles of nuclear fusion were posited in 1920,[22] only one design has thus far produced more power than it consumed. Since research into fusion power began in the 1940s, it took eight decades for researchers (including doctoral students) to obtain positive results and certainly a few more before society benefits from such an extraordinary breakthrough. For more than eighty years, much creativity was required for students to articulate the immediate impact of their work.

Indeed, the timeline for impact depends somewhat on where your research sits in the RDI spectrum. The mechanism for funding also depends on the type of research. For these two reasons, let us review two models of research classification.

BUSH'S MODEL

Research in engineering sits somewhere between basic and applied research, defined as:[23]

basic research: the search for new fundamental laws of nature; and
applied research: the development of specific solutions to targeted problems by applying fundamental results.

The closer you are to applied research, the easier it gets to define impact: it is easier to describe the output of a new instrument or system than it is to describe a new theory. Nonetheless, curiosity-driven research has its place in engineering schools. Curiosity leads to great PhD projects so long as researchers manage to fund such fundamental research projects. The first (modern) attempt at framing impact in the context of fundamental research came from Vanevar Bush, the US director of scientific research and development, reflecting on the impact of scientists in ending World War II. He described the linear model shown in figure 7.5 in which basic research is paving the way for applied research, development, and then innovation, respectively. The further to the left your research sits, the more imaginative you have to be to predict an applied outcome,

7.5 Bush's linear model for research and development leading to innovation.

but the more potential you have at influencing not just one subfield but an entire discipline. If all else fails, you may fall back on Bush's position and describe your fundamental project as "the pacemaker of all technological progress."[24] As the future expert on the topic, however, you are expected to be a little more specific and try to articulate how, if successful, your quest for knowledge may affect humanity. Remember, while you satisfy your curiosity, someone fights rain and mud to produce food. You owe it to society to come up with some explanation as to why you are researching this question and not another.

Bush's model is, however, imperfect. It gives the impression that knowledge only flows in one direction, from left to right, from academia to industry. History is full of examples of the industry driving the quest for knowledge. In his book *Pasteur's Quadrant,*[25] Donald E. Stokes describes a two-dimensional model plotting the quest for fundamental understanding against the considerations of use. Pasteur's quadrant shown in figure 7.6 contrasts:

Bohr's quadrant: research with a high requirement for fundamental understanding and a low consideration of use. Of course, the knowledge of Bohr's atomic model led to fantastic technological developments, but this is not the reason why Niels Bohr developed his model. His is purely curiosity-driven research;

Edison's quadrant: pure applied research is the kingdom of the inventor exemplified by Thomas Edison, relying on the fundamental understanding constructed by others to produce innovations; and

Pasteur's quadrant: research seeking fundamental understanding while having immediate repercussions on society is exemplified by Louis Pasteur's research. As a father of modern microbiology, his discovery of pasteurization was prompted by the milk and wine industries, yet required disproving the spontaneous generation theory, which was prevalent at the time.

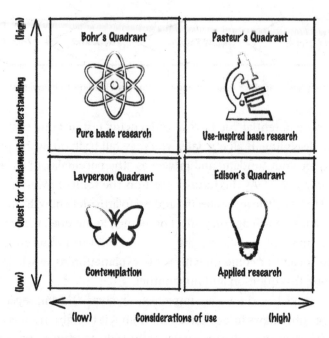

7.6 The quadrant model of scientific research.

The fourth quadrant represents the layperson (i.e., who may or may not have seen plenty of YouTube videos on the topic) who is observing natural elements (e.g., collecting butterflies, photographing nebulae) without the desire to advance fundamental knowledge or considerations of use.

PASTEUR'S QUADRANT

Pasteur's quadrant shows relationships between different types of engineering projects. The easiest project to frame in terms of originality and impact probably stands within Pasteur's quadrant. This is not to say that all doctorates in engineering must fall in that section. If your project falls within Edison's quadrant, its impact is somewhat easy to justify, but novelty is not. Conversely, in Bohr's quadrant, novelty is easy, while impact requires some creativity.

Missing from Stokes's model is the notion of fun. As a student, you benefit from finding a research project in the quadrant that best matches your idea of enjoyable research. It is immediately apparent to anyone who has seen even the pilot episode of Columbia Broadcasting System (CBS)'s *The Big Bang Theory*[26] that Dr. Sheldon Cooper would not be his happiest self in anything but Bohr's quadrant. Similarly, we could argue that Howard Wolowitz (aka, *Mr.* Wolowitz) belongs to Edison's quadrant. Neither would consider interchanging their positions, although great comedy came out of them trying to collaborate. Finding a project within your ideal quadrant is worth the extra effort needed to identify novelty and impact. Fun is indeed a key ingredient of breakthrough. Why did Henri Becquerel develop films that stayed in a drawer? The uranium salts he was testing had not been exposed to sunlight, and the films should not have been exposed to their "phosphorescence."[27] Perhaps he enjoyed the process? Serendipity is not a strategy, but it is undeniably part of the research process.[28] Indeed, as Louis Pasteur once said: "in the fields of observations, chance favors the prepared minds."

Box 7.6
Scientific Research

Place the following R&D endeavors in the quadrant model of figure 7.6:

- laser interferometer gravitational-wave observatory (LIGO)'s first observation of a gravitational wave;
- Steve Jobs's development of the iPhone;
- Stephen Hawking's black hole theory; and
- your own PhD project.

How would you have justified their respective originality and impact if you had proposed such a project?

SOURCES OF IMPACT

The easiest way to find impact is when your research is on the right half of figure. 7.6. A new biomedical instrument decreases mortality and morbidity, a more efficient code improves productivity, and a more efficient energy conversion process contributes to saving the planet. When your project is on the upper left side of figure. 7.6, the output is entirely curiosity-driven. But what about the method? Are you building a new instrument in your quest to understand black holes? Are you deploying new statistical methods that could be applied elsewhere? The exercise may be hard, but the reward is high if, indeed, your research sweet spot lies within Bohr's quadrant. However, if your research lies in the layperson's quadrant, run! When you come back, devise a research question that advances knowledge or improves humankind's conditions, or both.

Box 7.7

Emergence of Your Project—Part 2

Discuss the triggers of your research project and highlight the following:

- your research question;
- your hypothesis; and
- the underlying assumptions.

Additionally, write down the general and specific objectives of your research. As mentioned in chapter 4, consider using action verbs (i.e., "study" is poorly actionable, whereas "link X to Y" suggests a path and a measurable outcome), which help in defining impact. Finally, highlight the short-, mid-, and long-term benefits of your research project to society.

When looking for the potential impact of your research project, consider broadening your literature search to include industrial literature and white papers.

Definition 7.1: White Paper   *A report written by official instances such as organizations and governments, and, sometimes, industries to present an issue and solutions to a complex problem.*

If you are still having difficulty articulating what the impact of your project is, consider the governing pillars of sustainable engineering as a source of inspiration.

7.4 SUSTAINABLE ENGINEERING

Earth Overshoot Day marks the date when our yearly consumption of natural resources has exceeded what the planet can regenerate in a year. In 2022, it occurred just after mid-year, on July 28.[29] As engineers, the solutions we design must limit the negative impact we have on natural

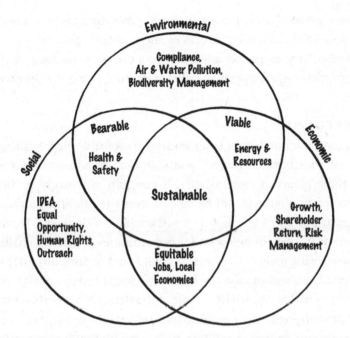

7.7 A Venn diagram of sustainable development. Socioeconomic considerations make the solution equitable, eco-economic considerations make the project viable, and socio-environmental considerations make the activity bearable. When all three pillars are considered, the development is sustainable. IDEA: inclusion, diversity, equity, and accessibility.

resources, and respect the principles of sustainable development. Indeed, many professional societies require that engineers pledge to find solutions respecting the best practices of sustainable development.

Definition 7.2: Sustainable Development *Defined in 1987 by the United Nations (UN), sustainable development concerns the ensemble of practices that meet the needs of a generation without compromising the ability of future generations to meet theirs.*[30] *The Earth Summit in 1992 in Rio da Janeiro offered an operational definition of sustainability: development that considers economic, social, and environmental impacts.*

The Venn diagram of figure 7.7 provides potential sources of impact for your research project based on three pillars:

Environmental pillar: Does your project limit damages to biodiversity and air and water pollution?

Economic pillar: Is your project improving a return on investment for shareholders or creating consistent and profitable growth?

Social pillar: Is your project applicable to a diverse population, is it promoting the respect of human rights or righting a long-standing wrong?

IMPACT FOR ALL

If you have the privilege of being entirely responsible for the emergence phase of your research project, you may elect to address one, two, or even three pillars of sustainability. If you join an already established research team, you may still be able to steer the project in a direction that contributes to some aspect of sustainable development. For example, you may keep an eye open for opportunities to address challenges associated with inclusion, diversity, equity, and accessibility (IDEA). In chapter 12, we will discuss the IDEA movement in the context of representation within the science community. Here, we argue that a proper solution to an engineering problem must address the needs of most members of society. Indeed, a solution that is tailored only to a portion of the population is not sustainable. Examples of engineering solutions with room for improvement (and, thus, greater impact) include:

safety features: for decades, crash test dummies represented average males in height, weight, and anatomical features. As a result, women are more likely to die or be injured during car collisions;[31]

diagnostic devices: the pulse oximeter is the instrument placed at the fingertip (or at the wrist in wearable sensors) to measure oxygen saturation using light absorption curves of oxy- and deoxyhemoglobin.[32] The melanin molecules present in the skin, however, also affect light propagation. As a result, oxygen saturation is often overestimated for the Black, Indigenous, and people of color (BIPOC) community;[33] and

AI: algorithms designed predominantly by males may reflect socially rooted, unconscious gender biases and address primarily the needs of the same population.[34] For instance, for centuries (some even suggest millennia[35]), women have tracked their menstrual cycle. Yet, when Apple launched its *HealthKit* in 2014 as "a hub for *all* your iOS fitness tracking needs,"[36] tracking one's menstrual cycle was not on the menu.[37] Even a team of the world's most talented programmers missed a feature so blatantly obvious to half of the population.

Sustainable development is not just about the environment. It is about maximizing the uses of limited resources to find solutions that benefit most. When defining your research project, look for avenues that minimize our collective footprint. Even if your project is about finding the most efficient way to diagonalize a matrix: if it aims to program a machine learning algorithm, if only one percent more efficient at the scale of the planet, you have a positive impact.

RESEARCH'S CARBON FOOTPRINT

You may also examine the carbon footprint associated with research activities as a researcher. At PolyMtl, a quarter of our greenhouse gas emissions come from research-related business travel. Your work ethic also plays a role in sustainability. Which conference must you attend, and for which other meeting can you engage with peers remotely? Could you combine a symposium with some vacation (and save money on airfare too)? Not everyone must be an expert in sustainable development, but everyone must question their habits to find better options.

7.5 SUSTAINABLE MOTIVATION

"If you want to build a ship, don't drum up the men to gather wood, divide the work, and give orders. Instead, teach them to yearn for the vast and endless sea."

Antoine de St-Exupéry

In the emergence phase of your doctoral project, you are also building motivation. Several milestones of your journey may affect the blissful optimism from your first day as a graduate student. A key aspect to maintaining motivation—along with proper support from your advisor, proper funding, and a keen interest in the project—is the alignment of your values, vision, and mission with that of the lab.

Box 7.8

Objective, Values, Vision, and Mission

Are the objectives of your project aligned with your core competencies or the competencies you plan on acquiring during the first years of your doctoral studies? What are your values, and is there a good overlap with that of your mentors? Aside from your immediate project, what is the vision of your lab (i.e., what is the legacy your advisor wants to leave behind?)? And what is its mission, that is, its *raison d'être*? How about you? What do you want to be when you grow up, and is a PhD in your current lab a proper path to reaching that goal?

This personal exercise is useful in preparing some lifelines for instances when motivation plummets. Some objectives of the project may be less appealing to you. For instance, you may be a hardware person who cannot avoid the occasional coding of servo-control software. Finding the motivation to debug a code may be challenging if the exercise is not your forte. In such a case, consider the overarching mission of your lab to extract what joy may be found from working on crashing software.

OUTREACH

Motivation may also be found in extracurricular activities. Local chapters of professional societies are always looking for volunteers to host outreach activities. Evangelizing engineering and applied science to school kids or laypeople is excellent for (self-esteem and) generating impact. A new generation may envisage a career outside their immediate circle of influence through you. Volunteering for outreach activities may bring your motivation back. Indeed, discussing with non-experts forces you to focus on the (noble) big picture instead of some immediate (pedestrian) challenge.

8

PLANNING AND ORGANIZING

"Failing to plan is planning to fail."

Benjamin Franklin

8.1 WHY PLAN?

Many see planning as a boring administrative exercise imposed by funding agencies or, perhaps, your doctoral institution. Rather, planning is a personal exercise aiming at:

- maximizing the prospect of success: delivering results within a limited time frame and budget;
- structuring and organizing your ideas;
- securing resources to match the scope of your project; and
- anticipating risks, hurdles, and conflicts.

Planning is a complex mental exercise used in research to avoid confusion and wasting precious time that could otherwise be dedicated to actual research.

Remark 8.1: *Planning is not an exercise performed at the start of a project and abandoned as soon as the proposal is accepted, nor is it a strict road map followed blindly without flexibility or updates. Rather it is a dynamic process. A good plan evolves over time. The team must be open to modifying the project if necessary.*

In this chapter, we discuss standard tools used in project management such as the WBS and the Gantt chart and adapt them to the planning of a doctoral project. Finally, we conclude this chapter with some time and priority management techniques to maximize the odds of getting some sleep during your PhD.

8.2   WORK BREAKDOWN STRUCTURE

Planning seems trivial until you start shaping your own project. When you do, you realize that the Devil is indeed in the details. Once you have established the objectives and scope of your project, one of the most challenging exercises is estimating how long the project will last. The US Department of Defense (DOD) proposed a tool to solve this challenge by creating the WBS: a hierarchical tree structure in which a project is broken down into dimensions, themselves divided into intermediate steps leading to work package (WP). While never entirely trivial, estimating the duration of a WP is easier than gauging the entire project. Estimation will further get easier and more accurate as you gain experience in project realization and management. If the WBS is complete, estimating the project's duration becomes a matter of adding that of all work packages.

DIMENSIONS OF A PROJECT

Level 0 of a WBS consists of stating the project's main goal. In the case of a doctoral project, the main goal is the general objective defined in your thesis proposal. Level 1 splits the project into dimensions. Figure 8.1 shows a typical WBS of a doctoral project. In this example, the dimensions include the SOs, defined in the thesis proposal, all academic requirements, and project (and career) management. Tips for creating a WBS include:

completeness: to be effective, each level of a WBS must span 100 percent of the project, including its management, and, in the case of a doctoral project, academic requirements;

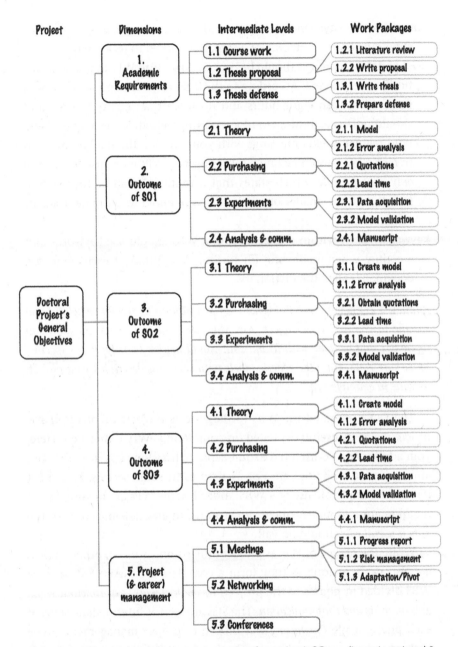

8.1 A generic work breakdown structure using the project's SOs as dimensions. Level 0 states the project's general objective. At Level 1, keywords represent each dimension. The words *Outcome of* should be replaced with project-specific keywords. Several intermediate levels are required before defining work packages. Dimensions do not necessarily have the same number of intermediate levels before reaching work packages.

outcome vs activity: when possible, and to avoid scope creep, the focus
should be on the outcome of a step, as opposed to its activity;

level of detail: a project should be broken down into WPs that may realistically be estimated. According to the 8/80 rule, work packages should
last no less than eight hours and no more than eighty hours,[1] and
should not exceed the span of a reporting period. For example, if you
hold regular weekly meetings with your advisor, the maximum duration of a WP should be one week. However, WPs must also follow the
only if useful rule, which states that activities should be further broken down into smaller bits only if it is required to estimate, track or
assign; and

keywords: to fit within a page or two, a WBS should use keywords and
a coding scheme allowing for further description of each item in a
document called the dictionary.

Remark 8.2: Planning Is a Personal Exercise   *Planning should be helpful and
specific to you and your project. Any student could copy the WBS shown in
figure 8.1 and be done with it, but that would neither be helpful nor specific.
Make sure you replace the generic labels with keywords describing your project,
as done in example 8.2.*

The DOD intended the WBS for large-scale projects involving many
protagonists: its usefulness relied on completing every dimension. Here,
your advisor and yourself have some sense of how long coursework takes,
and academic schedules are not as rigid as in some sectors. Figure 8.1
reflects the academic use of a WBS: indeed, some WPs are missing in the
requirement and management dimensions. In an academic context, the
only if useful rule should prevail.

While the choice of SOs as project dimensions is usual, other choices
are also valid, as long as they form a complete set. Figure 8.2 shows a
WBS divided in phases of the project: *theoretical work*, *experimental work*,
and *analysis and communication*. The SOs are seen as intermediate steps of
each phase, while *Purchasing* was moved to project management. Since
both sets allow for completeness, both are valid choices of dimensions.
Only the type of project will dictate which set is more appropriate, as is
illustrated in the following example.

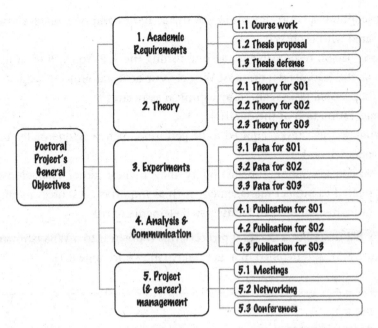

8.2   A WBS using project phases as dimensions. In some cases (see example discussed in box 8.2), breaking the project into phases is more useful than in SOs.

Box 8.1

Microgravity and the International Space Station

An experiment studying the effect of microgravity on cells requires sending specimens to the International Space Station (ISS). Even though many objectives may be specified in such a project, all theoretical modeling must be performed before sending cell cultures to the ISS, and all analysis done afterward. In such a case, defining a project according to phases makes more sense than using scientific questions or SOs. What are the dimensions of your research project?

DICTIONARY

The WBS outlines your project and should provide a comprehensive and uncluttered view of all WPs. A table—called the dictionary—provides a detailed description of each work package:

reference number: the unique identifier of the WP;

description: a short description of the activities and deliverables associated with each WP;

lead: person responsible for accomplishing the WP. You will most likely be the lead person on most WPs of your doctoral project,[2] aside from occasional tasks realized by technical associates;

time: estimation of the duration;

dependencies: tasks required to be performed before, sequentially, or in parallel; and

other considerations: the dictionary may include other characteristics related to project management such as important resources required, estimated budget, and anticipated elements of risk.

Box 8.2 describes a doctoral project broken down into a WBS (shown in figure 8.3), and detailed in a dictionary (shown in table 8.1).

Box 8.2

A Novel Microscope

A publication road map was presented in chapter 3 on the topic of clinical microscopy. Figure 3.6 showed the genesis of a novel microscope based on a new wavelength-swept laser, along with several other SOs omitted here for clarity. Figure 8.3 shows the associated WBS including the project's general objective in Level 0, the two SOs and a dimension on project and career management (academic requirements were also omitted). As much as possible, the WBS should describe outcomes instead of activities.

The dictionary for SO 1 of the example discussed in box 8.2 is presented in table 8.1. It includes a unique name or an identification number (ID) used in the WBS, a brief description of the WP, the initials of the person responsible for executing the WP (the lead person), an estimate of the duration (the time), and a column including other considerations related to project management. Other considerations include:

deliverables: the outcome of the activity, such as a report, plans, parts, an analysis or a recommendation;

criteria for completion: a metric that allows assessing to degree of completion of a WP;

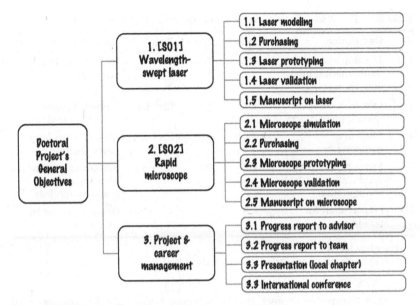

8.3   A WBS for the microscope project discussed in box 8.2.

resources: items required to perform a WP, including expertise, data, equipment, supplies, or documents;

budget: the estimation of the costs; and

risk: elements of uncertainty that may jeopardize completion.

A WBS may be used in and of itself as:

- a thought process tool to brainstorm on a project's activities with the team;
- an exercise that helps you structure a research project and allows you to adapt to (inevitable) changes;
- as an evaluation tool to highlight WPs over which you have little control and that may delay the project (such as purchasing);
- as a communication tool to assign WPs to other team members; and
- as a status reporting tool: some color codes or inset boxes indicate the completion percentage for each WP.

Each branch should be detailed enough to allow delegating it to another team member. But do not worry if distal WPs are not as detailed as proximal ones: a research project involves a fair amount of unknown. When

Table 8.1 The WBS dictionary for SOs 1 of the example discussed in box 8.2

ID	Description	Lead	Time	Dep.[a]	Other Considerations
1.1	Develop and validate mathematical model & python code	CB	3w		**Risk:** no analytical solution
1.2	Obtain & compare quotes; place orders	CB	2w	f2s[b]:1.1	**Budget:** $10,000 **Risk:** long lead time
1.3	Align laser components	Tech.	3w	f2s:1.2	**Risk:** requires laser safety training
1.4	Measure technical specifications of final configuration	CB	2w	f2s:1.3	**Resource:** frequency & optical spectrum analyzers (on loan)
1.5	Prepare draft for publication	CB	3w		**Delay:** revisions by coauthors

[a] Dependencies (see section 8.3).
[b] f2s: finish-to-start dependency (also in section 8.3).

planning a complex project, the WBS serves as the basis for another tool called a Gantt chart.

8.3 GANTT CHART

Project management requires tracking the completion of WPs and optimizing the order of execution that minimizes the project duration while maximizing the use of resources. A Gantt chart is a visual tool that lists all WPs and milestones in a vertical list. The horizontal axis represents time and shows their expected duration, start and finish dates, interdependencies, and progress.

Remark 8.3: *Specialized softwares, such as MS Project or OMNI Plan, generate Gantt charts with all bells and whistles required to lead multi-million-dollar development projects. For smaller-scale doctoral projects, free applications may be found online, including free LaTeX packages and simple spreadsheets. The tool you use is not as important as the actual planning and tracking of your project.*

The most straightforward Gantt chart is illustrated in figure 8.4, where activities are listed one after the other without any effort to optimize or parallelize the organization. Such a simple Gantt chart is useless. If your

INFORMATION CONTENT: ∅

8.4 When preparing a WBS, a Gantt diagram, or even an outline, be as specific as possible. If your Gantt chart could have easily been someone else's, modify it to reflect the specificity of your project.

planning concludes that performing one task after another is optimal, you do not need a Gantt chart. The tool is helpful when trying to identify which order minimizes the duration of a project and the critical path. For example, what can you do while you wait for equipment or comments from your advisor on a manuscript? Perhaps you may begin modeling the next phase of your project. Or write the cover letter for the editor, look up three to five references, and create an account in the journal's online system. With a Gantt chart, you can identify the main chain of events and satellite WPs that may be inserted into otherwise slow weeks.

Definition 8.1: Parkinson's Law *In the context of time management, Parkinson's Law states that work always expands to fill the time you allow for a task. So fight it by identifying the slow weeks and filling them with floating tasks, that is, tasks with no dependencies.*

DEPENDENCIES

You should determine the sequence in which WPs must be completed. Some depend on others: their relationships, or dependencies, have an impact on the duration of the project. Table 8.2 shows a list of tasks (from *A* to *J*) for which duration was estimated. The last column lists dependencies. Tasks *A* and *B* have no dependencies, but many tasks rely

Table 8.2 A dictionary for the Gantt chart shown in figure 8.5. At 50 percent effort, no more than two tasks may overlap at any given time

ID	Duration (weeks)	Dependencies	Effort
task A	5	—	50%
task B	4	—	50%
task C	3	f2s:B	50%
task D	7	f2s:A,C	50%
task E	5	f2s:A,C	50%
task F	2	f2s:B	50%
task G	4	f2s:D	50%
task H	3	f2s:E	50%
task I	5	s2s:H	50%
task J	2	f2s:G	50%

on their completion to start. Technically, four types of dependencies are possible:

finish-to-start (f2s): for a task to start, a prior task must be finished. For example, to begin an experiment, the lead time for equipment delivery must be over;

finish-to-finish (f2f): for a task to finish, a prior task must be finished. For instance, for you to finish a manuscript's submission, your advisor and all coauthors must have finished their revisions and signed off on the final version;

start-to-start (s2s): for a task to start, another task must have started. For example, for an intern to start a sub-project with you, you must have started your PhD yourself; and

start-to-finish (s2f): for a task to finish, another one must have started. For example, if you wish to hand over your experimental setup to another grad student, they must have started their PhD before you can finish yours.

The f2s is the most common dependency: I have managed projects for several decades without worrying too much about the other three. Understanding the broad concept of dependency is however critical in managing a research project and identifying its critical path in a Gantt chart.

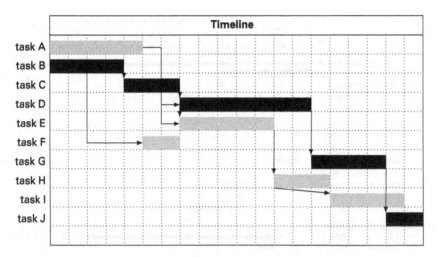

8.5 A simple Gantt chart based on tasks duration and dependencies listed in table 8.2. The highlighted bars show the critical path.

The tasks listed in table 8.2 are plotted in the Gantt chart of figure 8.5. Tasks are listed vertically and are followed by horizontal bars representing their duration (e.g., weeks in the current example). Tasks A and B have no dependencies. They mark the beginning of the project. Since all other tasks depend on them, they begin in parallel. At 50 percent effort, no more than two tasks may overlap. The tasks are thus placed to respect all durations and dependencies while minimizing the project's overall length.

CRITICAL PATH

In figure 8.5, tasks B, C, D, G, and J show an incompressible path known as the *critical path* (shown in black): their combined durations determine the duration of the project. Delaying any one of these items delays the whole project. Tasks shown in light gray have somewhat of a buffer, as fewer other tasks depend on them. Performing a Gantt chart analysis allows for identifying the critical path. Tasks belonging to this chain of events should be performed diligently with no obstruction.

Gantt charts may be used in doctoral research management. Indeed, figure 8.6 shows a very high-level Gantt chart based on the WBS of

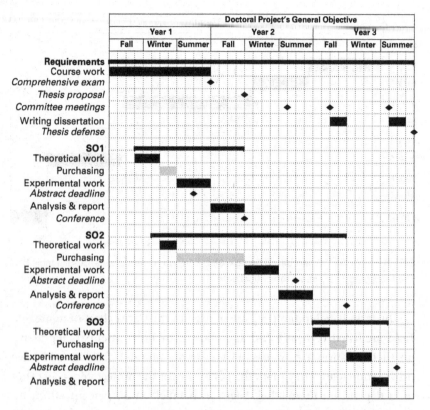

8.6 A GANTT diagram of a typical (read utopic) doctoral project. Tasks in black are mostly performed by the student, while tasks in gray are not, yet they are included in the chart as they affect the critical path.

figure 8.1. Work packages are grouped according to dimensions (Level 1), and important milestones are represented as lozenges instead of continuous bars. The timeline shows years, trimesters, and months. Such a high-level description is useful in a thesis proposal, but on a day-to-day basis, a more detailed, zoomed-in Gantt chart would allow for better progress tracking. Here, different shades of gray are used to distinguish tasks that belong to the student from tasks outside their control (e.g., lead times for pieces of equipment). This Gantt chart is based on 100 percent effort, which means that tasks that belong to the student should not overlap, except with coursework.[3] However, gaps created by events outside the student's control should overlap with student-led activities. The

Gantt chart shown here is a reminder to stagger activities from different SOs when possible to minimize the critical path.

Remark 8.4: Parallel Working *Gantt charts provide an occasion to explore occasions for working on several project dimensions in parallel. Most students approach their projects sequentially, which inevitably induces duration creep as they lose motivation, start procrastinating, and accumulate delays affecting the critical path. Working on several avenues simultaneously allows you to keep something moving at all times to maintain your momentum.*

Like most planning tools, the Gantt chart is not meant to be etched in stone and must be revisited periodically, especially as task lengths are guesstimated as best. I like using it at the beginning of a team meeting as a reminder of the project's complexity and the particular ensemble of tasks on the agenda. In the next chapter, we will use Gantt charts to track progress (and delays) and study risk management. Advanced Gantt charts may also account for contingency plans.

8.4 BUDGET

Let me start this section on budgeting by stating the obvious: you are not responsible for funding your doctoral project's research equipment or staff. However, in the spirit of training the new generation of R&D scientists, some schools ask that you include a financial forecast in your thesis proposal. Broad strokes typically suffice: your stipend and tuition, the salary of supporting staff, the use of core facilities, and the purchasing of small equipment and expensive consumables. For instance: several liters of liquid nitrogen consumed every week is considered expensive, but several boxes of Kimwipes® are not.

Table 8.3 shows an example of a budget for a doctoral project in engineering. Broad categories are used. A junior student may consult senior lab members to understand the order of magnitude for each budget item. An important conclusion from this budget is that despite the minimal wages paid to the doctoral student, the project costs society a significant amount of money. For example, in a project resulting in three publications, each article could cost taxpayers $100,000!

Table 8.3 A typical budget

Description	Unit	Qty	Unit Cost	Cost
Stipend	$/year	4	$24,000[a]	$96,000
Tuition	$/year	4	$5,000[a]	$20,000
Technician[b]	$/year	4	$20,000	$80,000
Use of core-facility[c]	$/year	4	$10,000	$40,000
Materials[d]	$/year	4	$5,000	$20,000
Consumables[e]	$/year	4	$3,000	$12,000
Conference	$/year	4	$3,000	$12,000
Publication	$/publication	3	$3,000	$9,000
Intern	$/year	2	$5,500	$11,000
			Total	$300,000

[a] Varies with countries and institutions.
[b] Assuming a yearly salary of $100,000, inclusive of benefits.
[c] E.g., one hundred hours/year of computer numerical control (CNC) machining.
[d] E.g., optomechanics such as lenses, lens holders, and posts.
[e] E.g., specialty optical fibers and components.

8.5 TIME AND PRIORITY MANAGEMENT

For Earthlings, a day lasts twenty-four hours. For some, these twenty-four hours are incredibly productive. In fact, many European PhD candidates finish their engineering doctorate in as few as three years! Given that the amount of time allotted to finish a PhD is finite, one must choose which tasks to prioritize and learn to work more efficiently. Here, we explore *priority management* techniques. We begin by asking ourselves: How does a PhD student spend time? One's typical schedule may be broken down into:

learning: taking classes, attending seminars and targeted training, studying for the comprehensive exam, and reading scientific literature;
researching: proposing, conducting, and communicating independent research;
training: assisting with supervising interns and teaching a class;
networking: participating in conferences and outreach activities, serving the scientific community;
living: maintaining a healthy and balanced lifestyle; and
procrastinating: *internet snacking*[4] and other activities not explicitly leading to one's growth or happiness performed as a way to escape important tasks.

> **Box 8.3**
> **Track Your Time**
>
> For an entire week, track how you spend your time and categorize your tasks. How are your activities distributed within these categories?

The question is: With so many activities, how best to manage priorities? Some would suggest the Pareto principle, which states that 80 percent of outcomes come from 20 percent of efforts. When applied to time management, the Pareto principle suggests that you focus on tasks that significantly impact your life. Of course, such an 80/20 rule does not apply to all categories listed above: obtaining significant research results requires a fair amount of brain and elbow grease. However, you might be able to use the Pareto principle in other spheres of your PhD, such as taking classes: work enough to get a decent grade, then hurry back to the lab.

THE EISENHOWER MATRIX

The Eisenhower matrix is named after an American general who served as a US president for two terms. He recognized that two categories of tasks landed on his plate: the urgent and the important ones. He is famous for having said: "I have two kinds of problems, the urgent and the important. The urgent are not important, and the important are never urgent" (Dwight D. Eisenhower[5]).

Even though, as scientists, we strive to only work on the important ("must cure cancer"), we, as humans, are conditioned to work on the urgent ("must finish this scholarship application"). The Mere Urgency Effect[6] describes our tendency to prioritize urgency over importance, even though the payoff of important tasks is more significant than the one for urgent matters. When facing a variety of urgent and/or important tasks, Steven Covey[7] suggests using a 2 × 2 matrix (shown in figure 8.7) to categorize tasks as:

important and not urgent—to schedule a task with no specific deadline that will bring you close to your goal. Need examples? Anything from performing a core experiment to brainstorming with colleagues to writing your dissertation;

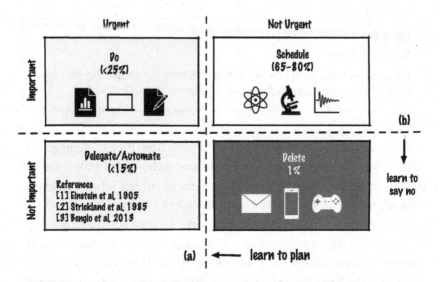

8.7 The Eisenhower contrasting, on the x-axis, urgent not urgent matters with, on the y-axis, important and not important problems. An important strategy in priority management is to move line (a) to the left and line (b) downward to maximize the time dedicated to solving important, yet not urgent, questions.

important and urgent—to do a task with an imminent deadline and consequences for not taking immediate action. Examples from doctoral life include submitting your thesis proposal and submitting a abstract to a conference;

urgent and not important—to delegate or automate a task that must be done rapidly but does not require your particular skill set. Delegating for PhD students might mean recruiting an intern to help acquire data or prepare samples. You may also delegate to a computer by automating tasks: preparing a bibliography, filtering data, or scheduling a meeting time with several faculty members. Never organize your references manually. Learn about bibliographic management software early on and delegate this task to an electronic friend; and

neither urgent nor important—to delete everything else from X notifications to Facebook posts to interrupting emails. You are not the servant of your computer; it should be the other way around. When working on important tasks, turn off your notifications. Set specific times to check emails and limit the time spent answering them. Remember that while haikus have been composed using X, I am unaware of any engineering PhD thesis fitting within its 280-character limit.

Figure 8.7 suggests an ideal distribution between quadrants. Time and priority management consists in maximizing the time you will spend working on important tasks at the proper pace while minimizing urgent matters and getting rid of neither urgent nor important tasks. Indeed, most urgent tasks are either the consequence of poor planning or are dictated by other people's needs.

Box 8.4
Track Your Activities

Make a list of all the tasks you accomplish in a given week, and place each item in an Eisenhower matrix. What can you do to reorganize your priority management to get closer to the optimal distribution?

Once you have filled your Eisenhower matrix, you may try to redistribute most tasks into the top left quadrant by:

moving line (a) to the left: by planning activities better, you minimize the number of activities accomplished under pressure; and

moving line (b) downward: by curating the activities you perform, you focus on those important to *your* project. Indeed, this means saying no to *some* people. *Some* being the operative word here: do not make the mistake of focusing exclusively on your PhD as collaborations are valuable and side-projects can be fruitful when properly balanced. However, by saying no to some opportunities, you say yes to *your* PhD.

TO-DO LISTS AND FROGS

Every researcher I know has a pad on their desk—often a legal one![8]— filled with outstanding tasks. The infamous **to-do list** is an essential tool for any high achiever. Unfortunately, to-do lists have a common flaw: they never seem to shrink. Not only do we tend to favor urgent over important but we also choose easy and fun over hard and ugly. Before long, our to-do list resembles a terror list that rapidly sparks more fear

8.8 First, eat that metaphorical frog. You will have the rest of your day to devote to more interesting tasks, without parasitic thoughts distracting you.

than joy. Once a week, offload the large items of your to-do list into your calendar. These are the important-and-not-urgent tasks. Moving them from the to-do list to the calendar will help allocate time and evaluate remaining resources. When a new opportunity arises, you are better equipped to accept or decline it. With experience, I find it is better to say no than to accept and then deliver results with a significant delay. An alternative to saying no is to reply: "I cannot now, but I could in three weeks." You let the other party decide *a priori* instead of delivering with a delay *a posteriori*.

With such a cleanup, my clutter-free to-do list only contains short tasks that I can easily slip in between meetings. Gone are the time-consuming tasks, but what about ugly ones? In his book, *Eat that frog!*, Brian Tracy[9] suggests that you should begin your day with your worst task, such as eating a frog (see figure 8.8), if such a task ever landed on your *plate*. Such a practice gives you a sense of accomplishment and frees your to-do list from ugly items that could act as parasitic thoughts keeping you from reaching your best state of concentration.

Remark 8.5: *(The Two-Minute Rule) Some tasks, however, are never worth postponing. Even if a task constitutes an interruption, if it only takes couple of minutes and is part of someone else's critical path, such as signing the*

8.9   The optimal way to fill the jar: rocks first, then pebbles, then sand. Then coffee.

authorship form for a joint publication, do it immediately. In an ideal world, though, all administrative tasks would be bunched together and presented to us only once a day (and in that same utopia, an elf would bring me coffee every hour, but I digress).

ROCKS, PEBBLES, AND SAND

Very little can be done to alter the passage of time. However, how could time be more efficiently spent?

Imagine a glass jar. In it, you place several rocks, as shown in figure 8.9. Is the jar full? It looks like it until you place a handful of pebbles into the jar. Several pebbles fit withing the interstices between the rocks, and now the jar looks fuller. That is, until you pour some sand in it. In such an order—rocks, pebbles, then sand—the jar contains a variety of stones. Each type finds its way into the jar. However, filling the jar in the opposite order—sand, pebbles, then rocks—prevents larger rocks from getting in. Figure 8.10 shows that the sand fills the bottom and does not fill the gaps between pebbles. This lack of optimization prevents larger rocks from getting in. In this rocky metaphor:

the jar is your calendar; better yet, the jar is your twenty-four-hour day;

rocks are activities taking a large chunk of time and requiring focus, such as performing a lab experiment, coding a simulation, drafting a section of a manuscript, studying for a comprehensive exam, and so on. Your family, too, counts as rocks: make sure you also schedule quality time with loved ones;

8.10 Sand, pebbles, and rock. No s. Adding sand first is a sure way for rock(s) to spill over.

pebbles require less time and focus but are essential elements of research, such as reading a scientific paper, rehearsing a presentation, comparing equipment for purchase, and so on; and

sand is somewhat mundane yet unavoidable tasks such as emails, organizing references, booking a hotel and transportation for a conference, and so on.

An efficient strategy in managing priorities consists of *filling* your calendar with important activities first. They require large chunks of your time and are essential to your graduation. After, you try fitting in pebble-like activities. Since they do not require large chunks of time, you can fit them in between your more important activities. Finally, in the small time slots remaining, you can fit in the sand: emails, proofreading a section of a manuscript, changing the font on that presentation slide, and so on.

Figure 8.10 shows another, albeit less productive, strategy in which you first add sand to your jar. In the opposite approach, you begin your day with emails, polishing every word, reviewing a slide deck for a conference presentation, tweaking every little detail, and wrapping your morning with social media because, well, by then, you deserve a break. Suddenly, it is lunchtime. In the afternoon, you tackle pebbles, and—before you know it!—there is one hour left before you can catch your last train to go home. Not enough time for a rock: a significant research activity, the foundation of your doctoral project, and the critical experiment is pushed back another day.

Use pebbles and sand to fill the cracks between rocks, and use the jar, or your calendar, efficiently. Once proficient in the technique, you may even fit in, between the sand pieces, coffee with a colleague. To make more room, select tasks that you can you may expedite, delegate, or even prune out.

Box 8.5
Highlight of the Day

In their book *Make Time: How to Focus on What Matters Every Day*, Knapp and Zeratsky suggest that you define a highlight for each day. Something urgent, something that brings you joy, or something that brings you satisfaction. Being deliberate in how you spend your time helps in staying motivated.

While answering emails is important and contributes to building your reputation as a reliable collaborator, no research has correlated zero-inbox policy with shorter PhDs. Without significant and original research results, you will not graduate. Make sure you allocate enough time for research first; then, between lab sessions, answer emails and book that flight for a conference.

8.6 WORK BETTER, NOT (JUST) MORE

Breaking your project into tasks is one thing; finding the right state of mind to accomplish these complex tasks daily is another. If you spend more time than anticipated on many tasks, perhaps reviewing some basic concepts associated with concentration is necessary. Concentration differs from attention. Attention is a reflex to a stimulus: a strong emotion, a physiological state (hunger, tiredness, temperature, pain), sensory variations (light, sound, odor), and, finally, events that appear fun, important, or valuable. Concentration is what shields the brain from taking its attention to a stimulus and allows you to focus on a complex problem or task. Concentration is not innate: it is acquired with practice while controlling the negative effect of internal and external factors. Internal factors include:

physical and psychological health: try to sleep and eat well (ideally not at
   your desk) and take time out to interact with friends and family;
stress: exacerbated by personal and complex problems, lack of time, and
   adaption to a new environment (see chapter 12);
parasitic ideas: from daydreaming to the grocery list, the mental load
   needs to be sorted out to allow for proper concentration. When parasitic ideas keep intruding during a work session, write them down and
   address them during your break; and
interest: perhaps you are not enthusiastic about all topics of your PhD.
   You cannot control mandatory topics, but you can control the method
   you use to study (perhaps join a study group with friends) and the value
   you attribute to the topic (perhaps it is a diving board for something
   more interesting down the road).

Even if your project is fun, some tasks may not be. You have two options.
You may trick your brain into being interested in turning a tedious task
into a valuable one, the equivalent to following one of Bocuse's recipes
to make the frog taste better.[10] You may also split an enormous task into
smaller ones to multiply the feeling of satisfaction from having completed
parts of it. If you still cannot fake enthusiasm for a task, get rid of it as
soon as possible. As mentioned earlier, if part of your day consists of doing
something as disgusting as eating a frog, make sure it is the first thing you
do in the morning. Otherwise, it will become one of those parasitic ideas
that clutter your mind and prevent you from reaching your highest level
of concentration.

Since there is no switch to turn concentration on, consider also working on external factors. Perhaps, like me in grad school, you have enough
roommates to assemble a small volleyball team. If so, consider booking a
library space or renting a desk at a co-working space (some even include
bottomless coffee, which, for some, offsets the rent). Work with the minimum number of electronic companions; turn off notifications while you
work on demanding tasks. Some even go as far as to create two accounts
on their laptops and exert parental control on the *serious* account to block
social media websites and applications. As a middle ground, consider separating your personal and institutional email accounts to allow you the
right to disconnect from work.

THE ZONE

Identify the moment of the day, or perhaps, the moment of the week in which you are the most productive. Schedule the most arduous task in that zone and do not let others interfere with it: try not to accept meetings at that moment, and do not let others interrupt you. If you have interns, make sure they are aware not to seek your attention at that moment. To alleviate your guilt, remember that limiting your mentoring to regular office hours allows your interns to learn independently and develop autonomy.

In psychology, being in the zone is a mental state known as the flow. Flow describes an immersive state in which both productivity and enjoyment are maximized. Figure 8.11 represents tasks that are susceptible to take you to your zone. These tasks are challenging and necessitate the set of skills in which you excel.[11] People who have experienced flow describe an intense focus, an intrinsic motivation, and a feeling of timelessness ("It is 2 a.m. already?"). Yet, some suggest that this task would last between sixty and ninety minutes: enough for flow to develop and just enough to justify a break.[12]

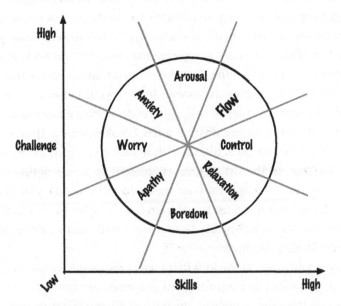

8.11   Flow. A mental state that may develop when the right challenge is presented to someone with the right skill set.

Flow needs indeed to develop, which implies some "setup" time to reach the proper mental disposition. When working in your zone, minimize interruptions, even short ones, as, every interruption will require you to redevelop flow. If it takes four minutes to recover from an eight-minute interruption (e.g., receiving and replying to an email), being interrupted five times is equivalent to having worked one less hour. Research found that interruptions further increase stress levels as people compensate by working under increased pressure and make more mistakes.[13]

Flow also requires that your goals are well established. When you feel overwhelmed with tasks, take a step back and remind yourself of short-, medium-, and long-term goals. Create a new list if you have forgotten what they are.

CROSS-PROCRASTINATION

Among my (many!) roommates was an Olympic athlete. While we were all trying to finish our respective PhDs, she was rowing to an Olympic silver medal. Very few experiences teach discipline like training—or even watching someone train—for the Olympics. What she also taught me was the notion of cross-training. In addition to rowing on machines and on the Charles River (while heroically avoiding Canadian geese), she would practice other sports to train complementary muscles and find yet different ways to boost her stamina. The idea of cross-training can be translated into what I call cross-procrastination. Indeed, your brain needs a break from intense activities such as debugging code, writing an article, or aligning optics. Perhaps, before wasting a few hours watching thirty-second videos, alternate intense activities with lighter ones, such as reading an article, looking up the list of speakers present at an upcoming conference, or tweaking a graph for your thesis. In other words, you may use light items on your to-do list to procrastinate on other intense activities.

Better yet, take a real break. Go out for a walk. Grab a coffee with a colleague. Stretch. Meditate.

Many would laugh at the idea that I may provide advice on time management. Too many last-minute grant proposals probably prompted me to reflect on the issue. Moreover, the birth of my son certainly influenced my priority management. If you have read this chapter thus far, I ask that you be indulgent with yourselves. You are new at this: both research and time/priority management. You may eventually fail at organizing the perfect jar—the ideal schedule—and may have to pull an all-nighter here or there to meet a deadline (or three deadlines at once). It happens to the best of us. What should not happen, however, is you relying on an ever-expanding jar, a permanently stretchable schedule. Not sleeping is not sustainable in the long run. Furthermore, you will find that the amount of rocks increases with time linearly (or exponentially) with seniority. Now is an excellent time to start developing strategies that allow you to transition from your work-work balance into a much more enjoyable work-life balance.

9

CONDUCTING AND ADAPTING

"The great tragedy of Science—the slaying of a beautiful hypothesis by an ugly fact."

Thomas Henry Huxley

After defining and planning, you are embarking on the execution phase of your project. This chapter discusses critical aspects of conducting research: proper communication with team members, risk assessment and mitigation, and ethical conduct. In addition to understanding these essential topics, remember to approach your research project with the proper attitude:[1] approach the implementation stage with enthusiasm, proceed with discipline, embrace all aspects of your work, even if—or, especially when—your project takes you outside of your comfort zone, and persevere in communicating. What? Persevere in communicating.

9.1 GETTING STARTED

A successful thesis proposal defense often marks the tipping point from planning to performing research. The kickoff meeting is an excellent opportunity to begin such a critical phase.

KICKOFF MEETING

The kickoff meeting indicates to all players the official beginning of the game. The general objectives of such a meeting are:[2]

contact: get in touch with all stakeholders, confirm that the main elements of planning are still pertinent and realistic, validate each other's interests, and identify common values;

project: present the project's objectives and context using the final version of your thesis proposal as the initial road map;

introspection: reflect on the questions and challenges you faced during your thesis proposal: What new insight has been gained? Should you revise certain elements?

protagonists: clarify roles, responsibilities, and expectations;

resources: verify the availability of resources; and

communication: establish communication preferences regarding the frequency of meetings and updates.

It is often hard to imagine, but some team members may be busier than you are. In biomedical engineering, for example, it is customary to include surgeons in the research team. Some students have had to scrub in to meet with medical team members between surgeries at hours that are better left unsaid. If a busier collaborator desires to meet on one platform instead of another, please consider this a lesser evil and acquiesce. Any platform is better than meeting before rounds! The kickoff meeting is an opportunity for stakeholders to agree on a match plan and the nature of each other's input and the project's output.

In engineering, projects result in several types of outputs, including scientific communications, IP, products, and services. A common concern from graduate students involves not publishing their work that results from an industrial collaboration for fear of revealing proprietary knowledge. The team should address such a concern early on and find mitigation strategies. Indeed, as a budding scientist, you must establish credibility as a researcher, and publications are an essential part of your portfolio. In addition, as a young engineer, you are also interested in creating innovations with a substantial impact on society, a route that involves some level of IP protection. The two concepts are not irreconcilable as long as all stakeholders acknowledge your needs as a graduate student and work toward a dissemination plan.

Remark 9.1: On What Terms? *A research agreement between your lab and an industrial partner or collaborator typically includes terms for publication, IP cessation, and requirements for data management. Someone from your school's research office may assist the team in finding a publication strategy agreeable to all. That person should also assist you in understanding the terms of an NDA or IP cessation agreement if you are asked to sign one.*

Authorship is more easily discussed early on, especially on collaborative projects involving several graduate students competing for the spotlight. A tentative publication road map is an excellent tool aiming at providing enough scientific material to satisfy the curiosity of all. Discussing authorship when the stakes are the lowest directly conflicts with Sayre's law, but it has worked relatively well throughout my career: "Academic politics is the most vicious and bitter form of politics because the stakes are so low" (Wallace Stanley Sayre).

Following the guidelines outlined in chapter 3 and in Patience et al.[3] allows you to guide the choice of authors with best practices in mind.

COMMUNICATION

A key aspect of conducting a project is developing successful communication strategies tailored to each player-message pair (see figure 9.1).

Box 9.1
Communication Means

Communication occurs through various means. Can you associate an appropriate medium with each pair shown in figure 9.1? Discuss your results with lab mates and perhaps with your advisor.

- one-on-one or group meeting, virtual or not
- hallway chat and elevator pitch
- email and snail mail
- text/instant messaging
- social media announcement
- blog post and wiki
- presentation

- conference proceeding
- publication, and so on

In some cases, there exist clear answers. For instance, research results are typically communicated to the scientific community through conference proceedings and peer-reviewed publications. In most cases, though, it is a matter of personal preference.

Some groups adopt a group communication platform for instant messaging, while some prefer formal emails. Some groups store information on an internal wiki, while others rely on people's memory. Inquire about what is currently in place, try to adapt to what is already used, and perhaps, participate in implementing a new strategy. Nevertheless, be sensitive to your advisor's reality when suggesting a new approach (see figure 9.2). Indeed, the COVID-19 pandemic has accelerated the deployment of new communication methods. I fondly remember the days when ICQ[4] was my unique instant messaging platform. In a typical week, I now employ upwards of twenty platforms, more if you include snail mail and people who insist on leaving voice messages. In a typical advisor fashion,

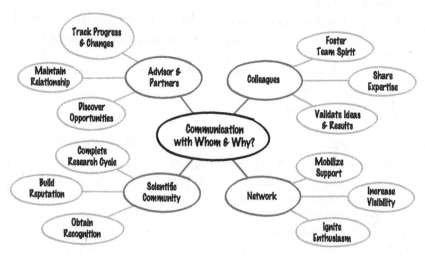

9.1   Communication: With whom and why?

9.2 Before suggesting a new communication method, please inquire as to what is already used in your lab.

I just expressed my preference through sarcasm. If you are immune to such a twisted communication technique, feel free to ask directly which method is preferred. Playing phone tag is not on my list of fun ways to get in touch.

What should you communicate with members of your team? Any news:

good news: let your team know when milestones are reached, research yields good results, scientific communications are accepted, or a scholarship, award, or grant is bestowed. Celebrating positive events contributes to preserving the morale of the team and your motivation throughout the project;

bad news: let your team know that if you experience some delays, research does not yield expected results, Reviewer 2 was more brutal than usual, or resources are running low—naming a problem is the first step to elaborating a solution. Incidents and near-misses that may have occurred are also worth sharing as they allow everyone to learn from a mistake and improve everyone's performance; and

ugly news: life throws curve balls at you, and it will not stop just because you are doing your PhD: family members will get sick, papers will get rejected, motivation will fluctuate, and instruments will break. While one does not typically broadcast ugly news on social media, you should have by now the ear of your advisor for such events.

9.3 Meetings should be used when dialogues are required and not for lengthy monologues.

"I broke an expensive piece of equipment by dropping another expensive piece on top of it."

My most creative graduate student

Remark 9.2: Mistakes vs Negligence   *Before using any lab equipment, ensure you have the proper training, first and foremost, for safety reasons and also to avoid breaking expensive instruments. When failures happen, however, your advisor must know, and you should assist in organizing repairs or re-orders. My point of view, as an advisor, is that the only way not to break anything is to not go to the lab. I would much rather replace broken pieces than not have results at all. Let's not, however, confuse occasional mistakes with gross negligence or repeated errors.*

No one likes leaving a meeting room thinking it could have well been an email (see figure 9.3). Meetings are useful when they encourage discussion: exchanging ideas to reach a decision. The HHMI proposes guidelines for effective meetings:[5]

agenda: solicit agenda items from key players, assemble the agenda, and distribute it before the meeting;

roles: assign clear roles such as introducing, moderating, and taking notes;

actions: for each item on the agenda, prepare discussion points, make a decision, and determine action items;

prepare: agree on a preliminary agenda for the following meeting; and

follow up: prepare a brief meeting summary, including a list of action items with their owner.

In a large group, try rotating roles, particularly when it comes to note-taking so that you avoid the cognitive overload associated with actively participating in the discussion while writing down the main resolutions. Beware not to systematically burden the same person with note-taking so you may take advantage of everyone's participation. Research has shown that women engineers are over-solicited for documenting meetings,[6,7] preventing their full involvement in the discussion and, eventually, pushing them out of technical roles.[8]

Remark 9.3: Electrons vs Emotions *Electronic mail does not convey emotion, intent, sarcasm, or cultural and societal codes well unless significant effort is made by the sender. It is easy to perceive an unpleasant tone in a message, even when none exists. Remember that we (advisors) are (typically) on your side. The same goes for (constructive) criticisms about your* **work**. *Remember that the* **work** *is being critiqued in an effort to elevate it. Indeed,* **you are not the one being criticized** *unless explicitly spelled out by the critic. Communication always involves a combination of information and emotion. Not all humans are skilled in the emotional component. #understatement*

9.2 ACTIVITY TRACKING

As the leader of your doctoral project, you are responsible for (making and) tracking progress to ensure timely graduation and avoid project scope creeping.

PROJECT PROGRESS

Organize meetings regularly, as opposed to when problems occur, to establish a good working relationship and, perhaps, avoid potential conflicts. A Gantt chart is a nice tool to visualize progress on all dimensions. Figure 9.4 shows how the partial coloring of the bars approximates the completion percentage. When tasks take longer than anticipated,

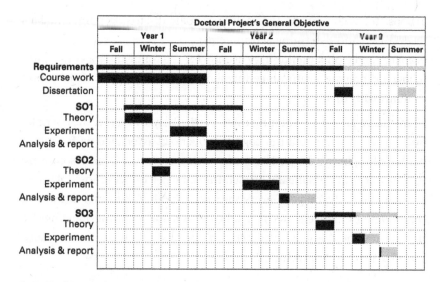

9.4 Tracking progress on a Gantt chart. Here, the gray portion of each bar represents unfinished work.

remember the project management triangle: you may modify the number of resources or the scope to comply with a deadline.

MOOD

On a scale from one to ten, how happy are you at the moment? is a question I regularly ask team members to assess their moods. Even as a project leader, you are not responsible for maintaining other people's good spirits, but you will benefit from identifying sources of friction early on and contributing solutions. Small problems are always easier to solve than large problems. Do not let small problems balloon out of proportion.

LITERATURE

Most students define the literature review as an initial phase of their doctoral studies, which is fair given that, at one point, one must submit their research proposal. The tricky aspect of placing the literature review as an item in a Gantt chart is that it implicitly states that one may be done reviewing the literature, which is never the case. Continuous screening of the literature is a perpetual activity for any researcher. Strategies to keep up with the ever-expanding body of published work include:

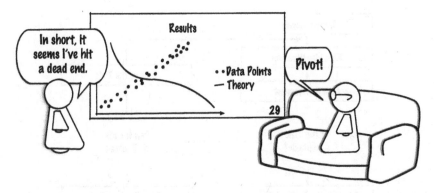

9.5 The only certainty of research is that you will have to adapt to change at some point. When you have to: pivot!

- setting up alerts by *following* competitors on several platforms;
- getting involved in the reviewing of upcoming articles;
- organizing a journal club with students working on related topics; and
- (procrastinating by) reading digest websites[9] to keep you abreast of an adjacent field that may relate to your work. You never know where that next great idea will come from—general scientific curiosity helps in that regard!

ADAPTING

You can modify the trajectory of a project as long as you do so with all protagonists informed and in the loop. Some elements you will control, and some elements you will not. Risk management techniques developed in section 9.3 will contribute to minimizing nasty surprises. It is necessary to acknowledge that your thesis proposal is not etched in stone. You can modify your objectives in light of dead ends or more promising avenues. You might even have to pivot entirely (see figure 9.5). As for many aspects of your PhD (your life?), communication is vital.

9.3 RISK MANAGEMENT

Part of conducting a research project involves dealing with avenues of uncertain outcomes. Some have a positive impact (i.e., opportunities), while others jeopardize the success of the operation (i.e., threats). This

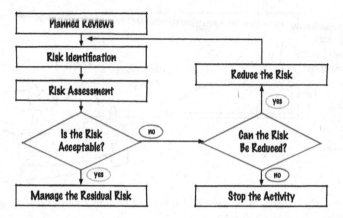

9.6 The risk management process.

section describes elements of risk management adapted to engineering research projects.

Definition 9.1: Risk *Risk is defined as the effect of uncertainty—potential events and their consequences—on successfully reaching the project's objectives.*[10]

Definition 9.1 breaks the notion of risk into individual components:

event: the situation affecting the course of the project, with internal (e.g., poor planning, wrong assumptions, bad foundations) or external (e.g., another lab publishing before you do) origins;
probability: the likelihood of such an event happening; and
impact: the severity of the consequences of such an event occurring.

Figure 9.6 shows the process of risk management involving identification, assessment, and mitigation.[11]

IDENTIFICATION
When asked what might affect the outcome of their doctoral research, first-year doctoral students often answer:

- "I fear not getting good results";
- "I fear being scooped";
- "I fear not being adequately supervised";
- "I fear not being able to publish due to IP considerations"; and
- "I fear running out of funding."

During group discussions with other students, several more risks emerge. One of the best ways to identify risks associated with a project is by brainstorming with team members.

Box 9.2
List of Risks

Make a list of the risks related to your research project. Then, brainstorm with your advisor and team members to create an exhaustive list.

When performing risk identification, consider these categories:

scientific or technological: poor results or performances stemming from inappropriate methodology, an invalid hypothesis, or experimental method;

managerial: inadequate project definition, preparation, leadership or adaptation: research management is a balancing act between lax planning resulting in vague objectives and rigid execution banning the exploration of emerging opportunities;

organizational: ineffective teamwork for reasons ranging from poor assignment of responsibilities to interpersonal conflicts (incompetence, personality traits, or availability);

financial or resource-related: deficient resource allocation (quantity, quality, and prioritization). For example, one may fail to put aside resources for repeat experiments to answer questions raised by Reviewer 2 (see figure 9.8). For graduate students, financial risks also include graduating past the scholarship funding period;

external: triggered by out-of-control elements, such as a publication from another lab on a similar topic, a change in regulation, or an industrial partner pivoting its core business or going bankrupt;

internal: excessive perfectionism—a trait often found in graduate students—involves refining one's thesis past the point of diminishing returns (aka *Gold Plating*) or not being able to wrap up (i.e., write and publish) an experiment until every aspect has been explored (aka *scope creep*); and

nature of the project: some projects propose incremental advancements, while others target disruptive research.

Low Risk / Low Reward   High Risk / High Reward

9.7 Zooming in on the frontier of knowledge. Brzutowski's Swiss cheese model shows a rough transition between the known and the unknown. Holes in the known sphere include incomplete theories, while bubbles outside the sphere represent data that the current theoretical framework cannot explain.

INCREMENTAL VS DISRUPTIVE RESEARCH

Even with the best planning, team, and intentions, not all projects are created equal regarding risk. It is the nature of research to venture into the unknown, which brings us back to the frontier of knowledge, discussed in chapter 2. While we previously described the boundary between the circle of knowledge and the unknown as a smooth one, Prof. Brzutowski argues that the boundary might resemble Swiss cheese, with holes and pockets on either side.[12] Figure 9.7 illustrates the Swiss cheese frontier of knowledge. Holes inside the sphere represent low-risk research that is incremental in nature.

Box 9.3

A Particular Solution to Laplace Equations

The potential flow model, which is represented by the Laplace equation, is well-known in the field of aerodynamics. Indeed, its numerical solution was developed in the 1960s.[13] Solving such equations for a particular wing configuration in fluid mechanics may be considered incremental. However, if applying its solutions paves the way for new aircraft technologies, it may be considered seminal work in aerospace engineering.

Research directly on the frontier of knowledge is moderately risky: quite a few researchers compete for these moderate-impact projects. Outside the sphere, researchers attempt to connect pockets of knowledge with current knowledge. Such projects are truly disruptive but hazardous: when successful, they affect the current thinking paradigm. This level of risk is inherent to the project. There is no right or wrong project type: as long as you are advancing science and the risk level is not too intimidating.

Box 9.4
Disruptive Research: The Photoelectric Effect

Observed before 1905, the photoelectric effect—the emission of photons when the matter is illuminated at specific wavelengths—contradicted the classical electromagnetic theory. Einstein and Planck reconciled this observation by proposing the photon as a quantum of light energy, ultimately leading to a paradigm change: quantum mechanics. Not every PhD, however, ends with a Nobel Prize. Very few do.

One of the risks associated with disruptive research is how lonely it is. Indeed, according to Prof. Brzutowski, most reviewers sit at the boundary and are ill-equipped to appreciate research far from what is known. Another risky aspect concerns researchers hitting dead ends. While negative results should be made public (to prevent others from wasting resources on the same topic), very few journals are dedicated to failed experiments. The discussion chapter of your dissertation is a nice, if not the only, place to discuss failed experiments.

HURDLES

The notion of uncertainty associated with risk identification is primordial. Events anticipated with 100 percent certainty and a high potential for deleterious effects are called hurdles. When uncovered, hurdles must be promptly attended to and downgraded without procrastination. There is no point in postponing dealing with an element that might cause a project to fail. In engineering research, hurdles are often associated with the scientific or technological dimensions or the availability of crucial resources. Scientific hurdles might inspire future research avenues.

What happens now that you have identified the risks associated with your project? Do you spend sleepless nights imagining the worst, or do

NUMBER OF REVISIONS: ▢ ▢ ▢ ▢   (and counting)

9.8 When sending reviews to the authors, editors typically place the most favorable review first, such that the reviewer with the most opposition become the (in)famous Reviewer 2.

you make contingency plans for every risk identified in this exercise? Enters risk assessment.

ASSESSMENT

Risk assessment begins with dissecting identified risks into these three components:

cause: the source of uncertainty;

event: the step of the project affected by such uncertainty; and

consequence: the consequence of an event being uncertain.

Box 9.5

Missing a Deadline—Part 1

The following risk is broken down into three components:

Late procurement delays data acquisition, resulting in missed deadline.
    cause   *event*   *consequence*

The risk matrix allows ranking risk by scoring for probability and impact.

PROBABILITY SCORE

For each risk, evaluate the likelihood of occurrence, p. To avoid analysis paralysis (i.e., "Is this 76% or 78%?"), use a global score:

1—not likely: $p \leq 30\%$;
2—likely: $30\% < p \leq 70\%$; or
3—highly likely: $70\% < p$.

IMPACT SCORE

Institutions such as NASA use standardized scales, ranging from discomfort to loss of life. Unless you plan on landing a human-crewed mission to Mars during your PhD, and since we do not expect your PhD project to result in loss of life, consider the following scale:

1—marginal: affects a minor aspect of the project;
2—critical: menaces one major aspect of the project; and
3—catastrophic: jeopardizes the entire doctoral endeavor.

Definition 9.2: Criticality   *Following NASA, we define criticality as the product of the probability of failure and the impact of failure:*

$$Criticality = Probability \times Impact. \qquad (9.2)$$

When probability and impact are scored using a one-to-three scale, the project manager scores the criticality on a scale ranging from one to nine, divided into low, moderate, and high.

The criticality acts as a threshold for deciding when to apply mitigation techniques:

high (or unacceptable): an event that is likely to occur and, if it did, would very negatively impact the course of the project should be addressed immediately to downgrade it to a lower criticality;

low (acceptable): an event that is unlikely to occur and, if it did, would only slightly impact the course of the project should not keep you from sleeping at night (i.e., ignore); and

moderate (acceptable with mitigation): for anything in between, consider a mitigation plan and monitor the situation.

The criticality is used in conjunction with a risk matrix.

		Impact		
		Low (1)	Moderate (2)	High (3)
Probability	High (3)	3	6	9
	Moderate (2)	2	4	6
	Low (1)	1	2	3

Urgency of Risk
Response Planning

■ High
■ Moderate
□ Low

9.9 The risk matrix was developed to determine the urgency in planning risk response.

RISK MATRIX

The risk matrix is a tool to help you keep track of the particular risks of your project and highlight which are critical and must be mitigated. Figure 9.9 shows a risk matrix acting as a legend for risk management. A high criticality (typically in red, here in dark gray) requires immediate action to downgrade the risk to a lesser criticality. Moderate criticality (typically in yellow, here in medium gray) necessitates elaborating mitigation strategies. Low criticality (typically in green, here in pale gray) must be monitored and addressed if it increases.

MITIGATION

Table 9.1 shows how a risk matrix is used to identify, monitor, and mitigate risks as part of managing a research project. The color code is based on the risk matrix (figure 9.9). The first column identifies risks: *getting scooped, confidentiality, unavailability of resources,* and so on. The following columns describe, respectively, the probability score, P, the impact score, I, and the criticality score, C.

When the criticality score is high (dark gray cells), immediate action is required to either eliminate the risk or downgrade it to a lesser criticality. Reducing P, I, or both are effective downgrade strategies. Risks highlighted in pale gray are moderately critical: mitigation solutions must be presented. Finally, risks shown in white are deemed lowly critical: no

Table 9.1 Risk matrix for a doctoral project in engineering

Risk	P[1]	I	C	Mitigation (for $C \geq 6$)
Getting scooped	3	2	6	Scout literature with automated notifications; devise a publication strategy that includes publishing the method
Industrial partner getting in the way of your publishing	3	3	9	Immediate action required: agree on a publication strategy with a timetable allowing IP protection in collaboration with your school's research office.
Unavailability of crucial resource	3	3	9	Immediate action required: make a plan to secure the resource through acquisition, collaboration, or a loan from a company.
Suffering from the impostor syndrome	3	2	6	Recognize that this comes with knowing what you do not know and find strategies to tame the beast.
Using dangerous equipment	$3 \to 1$	3	3	[Downgraded after proper training & safety protocol]
Advisor expecting a child[2]	3	2	6	Reorganize tasks to be autonomous during parental leave, work on a side project with a co-advisor, and establish a respectful communication method.[2]
Quitting your PhD	$2 \to 1$	2	2	[Downgraded since you are reading this book]
Advisor leaving or dying[3]	≈ 0	3	0	Identify a potential co-advisor

[1] P, I, and C are the respective scores for probability, impact, and criticality. This list is not exhaustive nor tailored to your situation.

[2] Discuss fear and expectations early. As a group leader, not knowing what is happening in the lab is worse than reading an occasional email. However, students should know better than to trust the judgment of a new parent a few weeks post-partum.

[3] Students always suggest this as a risk. I wish this was $P = 0$. Realistically, make a mental list of potential co-advisors.

mitigation measure is necessary, but such risks remain on the table for monitoring. The last column of table 9.1 presents mitigation strategies for moderate and high criticality. Such a table should be updated as the project unfolds: some risks arise, some risks vanish, and their scores fluctuate.

Box 9.6
Missing a Deadline—Part II

A mitigation strategy for the risk stated in example 9.3 could target the cause, the event, or the consequences:

Cause: Order material early, or borrow from another lab while you are ordering the replacement (i.e., reducing the probability).
Event: Is this data required for submitting an abstract (i.e., removing the event)?
Consequence: As a last resort, is there a post-deadline way to submit a contribution (i.e., attenuating the impact)?

Once a risk matrix has identified risks with a high criticality score, you should devise some mitigation strategies. Here are examples of common risks identified by students and possible mitigation techniques:

lack of knowledge: take classes, register for workshops, or leverage from collaborators' complementary expertise;
limited availability of resources: carefully schedule experiments with other users of critical resources, scout the area for similar instruments, inquire for loans from industries;
small budget: be on the lookout for scholarships and grants, assist your advisor in grant writing, and maximize your output from available resources;
tight schedule: avoid procrastination, evaluate the benefit of training an intern for some tasks, prune tasks that are not directly contributing to the project, communicate expectations better (e.g., create an internal calendar to avoid submitting to your advisor all conference abstracts from the lab the night before the due date); and
creeping scope: hold periodic review meetings to refine the objectives and keep an eye on the goal.

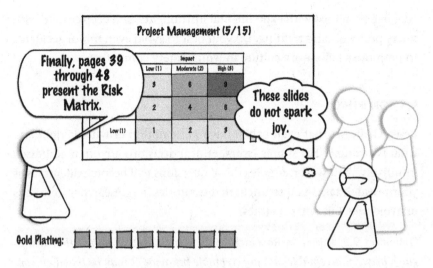

9.10 Risk management must be performed as efficiently as any aspect of project management. When presenting reports, charts, and graphs, prune out all elements that are not strictly relevant.

Refrain from cluttering a risk matrix with obvious and low-criticality items. In such a list, the Marie Kondo[14] approach is your friend (as hinted at in figure 9.10): thinking of what could go wrong rarely sparks joy, but knowing that a solution exists might.

Box 9.7
On Mitigating Risks

In your previous projects, have you had to deal with mitigating risks? If so, what risks were identified, and what corrective actions were taken?

BENEFITS FROM RISK MANAGEMENT

Anxiety-prone graduate students may prefer to avoid thinking about risk altogether, correlating ignorance with bliss. However, when it comes to risk, I believe forewarned is forearmed; in other words, early identifying possible failure mechanisms allows more time to design corrective measures.

Managing risk implicitly includes laboratory incidents and accident prevention. Never engage in dangerous activities, make sure you are adequately trained with experimental instruments, and keep current with safety protocols and training. Do your best to avoid even minor accidents to promote a laboratory culture in which safety comes first.

9.4 ON ETHICS

Research shows that even babies have some notion of what is right and what is wrong.[15] For simple issues, ethical decisions are often addressed intuitively.[16] However, simple ethical questions will become dilemmas as your research takes you to uncharted territories. For researchers, ethics is unavoidable. But what is ethics?

Definition 9.3: Ethics in Research *Ethics is the set of norms that distinguish between acceptable and unacceptable behavior. It may be based on your convictions (virtue ethics) or a code of good practice (deontology).*

Box 9.8
Your Ethical Dilemmas

Before reading the rest of this section, try identifying potential ethical questions from your research project. Ethical dilemmas can arise from all sections of your thesis proposal: objectives, materials and methods, data acquisition, analysis, and publication.

Before assessing the ethics of a situation, make sure you review the code of conduct of your school and that of the funding agency.

PRINCIPLES AND GUIDELINES OF ETHICS IN RESEARCH

Research is essential to the development of society. However, knowledge advancement should never precede the well-being and integrity of individuals and their communities. When evaluating the ethics of a situation, priority should be given to the following principles:[17]

respect for persons, animals, and the environment: the ethics committee of your school oversees research involving humans, animals, or cells. As an engineer, your deontology code further requires that your work

9.11 The honest and rational use of public funds is one of the pillars of ethics in research.

be in line with the best practice of sustainable development (defined in chapter 7);

the well-being of the community: this includes the safety of the researchers and general population, as well as the prompt communication of scientific results and, if applicable, the protection of IP resulting from the research project. The well-being of the scientific community also means that you must not plagiarize the work of others; and

the honest and rational use of public funds: as a researcher, you must conduct your research with integrity, which includes publishing all your data (in opposition to cherry-picking data points that fit the trend), citing the relevant work (and not just your own work to increase your h-index), and use public funds to advance research (and not as a way to reach fancy destinations, despite what figure 9.11 suggests).

ETHICS COMMITTEE

Universities usually have ethics committees, also called institutional review board (IRB)s, overseeing research with humans, animals, and cells. The committee provides guidelines for researchers and assesses the scientific validity and conformity of research proposals with current regulations and best practices.

RESEARCH WITH HUMAN SUBJECTS

From questionnaires to testing new medical imaging tools or pharmaceuticals, research on human subjects must be approved by the IRBs of all institutions involved in the research trial: the engineering school and the medical school. The IRB is the authority responsible for assessing the ethics of research projects involving human subjects. Based on the Belmont Report,[18] the ethical principles and guidelines for research with human subjects should be based on these three pillars:

respect for persons: people are autonomous and capable of choosing for themselves unless, due to various reasons, they cannot, in which case they require protection. Vulnerable people include children and prisoners;

beneficence: best effort should be taken to secure the well-being of people by first not harm, or, at least, maximizing possible benefits while minimizing possible harms; and

justice: people who benefit from research should share its burden, and vice versa.

It is the PI's responsibility to make sure research involving humans (or animals) receives the approbation from the IRB *before* the research project begins. Therefore, students should be involved in the accreditation process to abide by the methodology approved by the IRB and flag any new elements as they arise. You should also arrange for your work to be reviewed by the IRB. If the IRB only meets once a month, allow for a minimum of eight weeks to account for revisions.

BEHAVIORAL ETHICS

Behavioral ethics suggests that you can create a sense of how ethical a situation is by applying the following tests:

the legal test: Is this legal? Before attempting to assess a situation's morality, ensure you are on the right side of the law. If you are, move on to the next question;

the smell test: How do you feel about the situation? What does your gut tell you? Does it smell of corruption? Does it make you feel uneasy?;

the front page test: How would you feel if this personal decision you are considering regarding this situation was suddenly made public—could you justify your actions? Are you convinced enough to convince others?; and

the mom test: What would your mother think about this? What would she do? Could you justify your actions to her? If something bad happened, could you tell your mother that every mitigation that could be put in place was in place or that "you did not have the resources to do it properly but did it anyway"?

Box 9.9
Ethical Dilemmas in Engineering

Examples of research dilemmas encountered in engineering research projects include:

1. Is it justified to use animals or humans to validate my new medical instrument?
2. Can I use data from social media to feed my AI algorithm?
3. Can I use this pirated software just this one time to perform a complex simulation?
4. Should I use public funding to attend this scientific congress in Hawaii?

Apply the four tests to each situation and see which ones are morally justified. Hint: the third one fails the legal test—the answer is no, it is not legal and therefore must not be done. The answer to number four depends on whether or not the conference is legitimate: similarly to predatory publishing, phony conferences also exist.

CONFLICT OF INTEREST

A conflict of interest occurs when the personal interests of one person may interfere with the integrity of their professional decisions. Personal interests leading to conflict of interest include:

family: for example, the daughter of a professor takes their class;

pecuniary: for example, a researcher negotiates licensing rights to a technology developed jointly with their students; and

career advancement: for example, a professor must decide whether to publish or not the results of industrial collaboration.

The first reflex associated with conflicts of interest is to avoid them. However, the examples cited previously show that this is impractical:

Why should family members be forbidden to attend the same school? Researchers from licensing their own work? Or professors from interacting with industry leaders? The key is to disclose and manage conflict of interest. A quick brainstorm will reveal that: an independent grader could score the daughter's work; an independent committee could negotiate the terms of the license on the student's behalf; and early agreement with the school's research office could help decide what is published and when to allow commercial partners to protect any relevant IP appropriately.

AUTHORSHIP

Section 3.5 discussed what constitutes authorship in research communications. From your point of view, as the lead author of a communication, adding coauthors to your work acknowledges their participation. From their point of view, it constitutes a signature, read an approbation of what you wrote. Ethically speaking, you must present your document to all coauthors and allow for enough time for all to review the text, the poster, or a video recording before submitting your work to a journal, a conference, or even a local student-only event. If someone asks for significant revisions, make sure, when an agreement is reached, that all coauthors receive the final document—perhaps with an email thanking everyone for the fruitful collaboration.

PLAGIARISM

It used to be that the three most dangerous computer keys were CTRL, ALT, and DEL, which would instantly reboot your system (without saving your work). Nowadays, in academia, the most dangerous keys are CTRL-c and CTRL-v, which, when used in the right order, insert copied text into one's manuscript. A simple act that could get you expelled from your academic program, or ruin your career, on account of plagiarism.

To plagiarize is to "steal and pass off (the ideas or words of another) as one's own: use (another's production) without crediting the source" (Merriam-Webster[19]).

Plagiarism is often detected, including in written assignments, manuscripts, or dissertations. Detection, which previously relied on the erudition of reviewers, now exploits sophisticated algorithms to compare a new document to anything available online.

Strategies to avoid plagiarism include:

- reading guidelines from institutions (your school, publishers, granting agencies) to familiarize yourself with proper crediting of authorship (i.e., when to paraphrase, use quotation marks, citations, and so on)[20];
- when taking notes, inserting copied text in another color until you properly format (paraphrase or quote) the text to credit the original author(s);
- indicating the origin of visual support such as tables, figures, artwork, and so on. Obtain permission to reproduce material under copyright; and
- run your text through freely available plagiarism detecting tools as a last verification.

Beware of the transfer of copyright. When you publish a document, the copyright of your work is often transferred to the publisher. You may have to ask permission to reproduce figures in your dissertation. Worry not, as permission from legitimate publishers to authors of some work is almost always granted. But this does not mean one should not ask.

Box 9.10
Small World

Years before the dawn of plagiarism-detection algorithms, I was asked by a journal to peer-review a manuscript. The topic was close to that of my dissertation. The equations felt familiar; the text did too. Perhaps too much so. After a few minutes, I realized several paragraphs had been copied and pasted from ... one of my papers. How unlucky for the researcher to be evaluated by the author of the very paper they had plagiarized? Very unfortunate, especially given the meager number of papers I had published at that point! The scientific community is a small world. Do not rely on luck and review best citation practices.

10

CONCLUDING AND SUBMITTING

"Perfectionism is just fear in fancy shoes and a mink coat."

Elizabeth Gilbert[1]

When I left home to start my PhD, my undergraduate research advisor gave me a copy of Mary Schmich's pseudo-commencement speech[2] and told me this: "There are two kinds of theses: the perfect one and the finished one. Only the one you submit might get you a diploma" (Dr. Christian Moisan).

I smiled, thanked him, ignored his advice, and aimed for perfection. There is no need to actually download my thesis; I will readily admit that the one I produced was of the second kind. Still, I left MIT with a PhD, a job, a budding portfolio of publications, and a not-so-original piece of advice for my aspiring PhD students.

In this chapter, I explore the purpose and virtues of the dissertation in hopes of elucidating this final step in your journey. I also describe the hidden aspects of the thesis defense. Finally, I explore the process of concluding an engineering project with a documentation phase that often includes protecting IP.

10.1   THESES AND *ANNA KARENINA*

In this last phase of your PhD, you are required to write and defend a thesis that includes your significant contributions to the sphere of knowledge, as well as a critique of your work with respect to that of others. As a doctoral student, not only are you documenting your scientific journey through conference presentations and publications, but you also need to convince a jury of experts that you deserve the ultimate academic title. Unfortunately, even after several years of research, what constitutes a great—or even an acceptable—dissertation often remains nebulous for doctoral candidates. Tolstoy opened his famous novel *Anna Karenina* with this line:

Happy families are all alike; every unhappy family is unhappy in its own way. (Leo Tolstoy's Anna Karenina)

The same can be said about doctoral dissertations:

Outstanding dissertations are all alike; every mediocre dissertation is mediocre in its own way. (Yours truly)

Outstanding dissertations score high in every aspect. Mediocre ones either miss out on one (or several) aspects: impact, novelty, writing style, analysis, technique, or professionalism. First, let us review the purpose of a PhD thesis, then discuss its qualities.

PURPOSE

The doctoral thesis is the official vessel to disseminate the new knowledge you uncovered in the process of getting your PhD. The purpose of an engineering dissertation is to provide evidence of:

mastery of the field: the dissertation reveals the scholarship achieved during graduate school, surveys the current state of the art, identifies problems, and generates questions;

technical skills: it exposes the research strategy and the methods used to acquire novel and impactful results;

analytical skills: the document explains data analysis, thoughtful discussions, and publishable results;

autonomy: shows the student's global perspective on the problem to further contributions to the field;

writing skills: the manuscript documents with a clear narrative framework the student's research and thinking process, communicates it to the world, and provides a tutorial for the next generation of students; and

professional skills: the thesis exposes research performed according to a proper code of conduct and ethical guidelines.

SUCCESS CRITERIA

A doctoral dissertation is the ultimate academic deliverable, yet students are rarely given a correction grid as to what is expected from them. In other words, what are the standards used to evaluate dissertations? A survey of more than two hundred faculty members across several disciplines revealed that:[3]

[Faculty members] often make holistic judgments about the quality of a dissertation after they have read it. They do not have a mental checklist of items against which they assess a dissertation. (Barbara Lovitts)

The same group of faculty members, Lovitts reports, however, achieved a high degree of consistency when evaluating dissertations across four levels:

outstanding: described as *once in a decade*, the outstanding dissertation *display[s] a richness of thought and insight and make[s] an important breakthrough*. The question is original, its impact is felt in other fields, the results are significant, and their analysis is performed using novel tools. The student shows a mastery of the field that exceeds expectations, and the document is a page-turner. Figure 10.1 shows the outstanding thesis ranking the highest in every direction. Outstanding theses often reflect a rich interaction between student and advisor;

very good: while well organized and written, the very good thesis lacks a little *je-ne-sais-quoi* in some of the dimensions (an example is shown in figure 10.1). For example, the writing may be flawless without demonstrating the elegance of the outstanding thesis. In the example shown in figure 10.1, the topic is original, but the significance is not as great. The student masters the field, the analysis and results are strong, but the student may have missed opportunities to explore interesting avenues and connections completely;

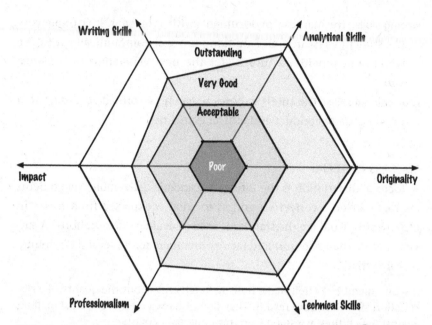

10.1 Radar plot of the qualities of outstanding, very good, acceptable, and unacceptable dissertations.

acceptable: the group of faculty describes the acceptable thesis as *not very*: not very original, not very significant, and so on. Figure 10.1 shows that the acceptable thesis is neither very original nor quite significant nor well written. It demonstrates technical competence without showing surprising results or sophisticated analysis. The acceptable dissertation nonetheless makes a small contribution to the field and grants its author a PhD degree; and

unacceptable: the unacceptable dissertation is, fortunately, a rare bird, as candidates often quit before reaching the dissertation stage. Such a document is described as unoriginal, insignificant, and lacking depth.

Box 10.1
Grading a Dissertation

Reading dissertations from previous students in your laboratory is a fantastic way to get a feel for how ideas are presented and discussed. As you are reading it, ask yourself how the dissertation scores using the radar chart of figure 10.1.

10.2 TYPES OF DISSERTATIONS

In addition to the traditional manuscript, some institutions allow for cumulative (article-based or hybrid) theses. Here, we paint the portrait of several thesis types in broad strokes: seek out your friendly school librarian or administrative assistant for details. Most schools also provide a template, which is a great way to ensure that format requirements are met.

TRADITIONAL DISSERTATION

A traditional (chapter-based) dissertation is written *de novo* as one coherent and comprehensive scientific manuscript. Typical sections of a traditional thesis include:

Front matter: in addition to the title and signature page, the front matter includes:

dedication: optional, yet an easy way to make your parents, partner, or even your cat happy;

acknowledgments: optional, yet an excellent way to make your advisor, lab members, and friends outside the lab—if applicable—happy. However, if you cannot find anything nice to say, refrain from using this section to vent, as printed words may outlast your feelings;

abstract: some institutions require one abstract in the school's official language and another in English, science's *lingua franca*; and

tables and lists: table of contents and lists of tables, figures, symbols, abbreviations, and appendices.

Main matter: also called the body of the thesis, it includes the following:

introduction: sets the scene, highlights the triggers for your work, and exposes the document's organization and student's contributions (i.e., a list of publications, conference presentations, and patents);

literature review: provides background and explains where your study fits in relation to recent work by surveying and critiquing the work of others to identify the knowledge or technological gaps you wish to fill. The literature review behind your research question and objectives culminates into the thesis statement (i.e., the research question, hypothesis, general objective, specific objectives, and assumptions);

materials and methods: exposes the study design, justifies the implementation, and explains the statistical analysis behind error estimation;
results: analysis and interpretation;
discussion: critique of your work concerning your original hypothesis
   and objectives, as well as that of competing groups, presentation of
   inconclusive research avenues, and/or paths not traveled; and
conclusion: summary of your contributions to advancing knowledge,
   comments on strengths and shortcomings of the research work, recommendations for researchers following in your footsteps, and outlook
   for the field.

Back matter: contains references and appendices.

CUMULATIVE (OR ARTICLE-BASED) DISSERTATION

The dissertation may be presented through a carefully curated selection
of published (or in some instances, submitted) articles or conference proceedings. Refer to your school's guidelines to verify what is accepted.
Since journal articles are extremely concise documents, several chapters
are added before and after those containing the published work to situate
the work within your broader doctoral project, adequately describe your
methodology, and provide a global conclusion on the work. In addition,
some paragraphs are added at the beginning of each chapter to reinforce
the narrative framework and the coherence with the other chapters and
make the student's contribution to the manuscript explicit. As the ability to write a dissertation is critical for PhD candidates, most institutions
require that the student be the first (or co-first) author of the publication
to allow inclusion within the dissertation.

In a manuscript-based thesis, the front and back matters do not differ
much from the traditional thesis, but the core of the thesis does.

Body of the thesis:

introduction: in addition to setting the scene and detailing the thesis statement, the introduction must highlight the common thread
   between the student's several publications. Here, the thesis organization must highlight both the student's contributions and the logical
   progression from one topic to the next;
literature review: same as for the traditional thesis;

first publication: introduce your publication by providing the context of
the work and highlighting your contribution to the publication. All
sections of the published manuscript are reproduced, including tables,
figures, and references, which are integrated into the lists of figures and
tables in the front matter, and reference section in the back matter.
Permission must be obtained from the publisher of your manuscript
when the copyright was transferred for publication;

other publications: idem;

last publication: rinse and repeat—very prolific authors are expected to
determine which papers merit inclusion in the dissertation (papers
contributing to the narrative and supporting the thesis); others go to
the appendices or are not included at all;

general discussion: while each paper already includes a discussion
section, here, you must provide a general critique of your entire thesis
concerning your original hypothesis and objectives, as well as that of
competing groups. The general discussion is also the chapter in which
you describe inconclusive avenues; and

conclusion: summary of your contributions, an analysis of research material with respect to the field (unless already included in the main
chapters), comments on strengths and shortcomings of the research
work, recommendations for researchers following in your footsteps,
and outlook for the field;

It may seem that the energy you save in not rewriting your work you
spend contextualizing it to obtain a coherent body of work. Writing an
article-based thesis saves you some effort as you write gradually. You are
not scrambling for data several years after performing the experiment or
realizing that you have not acquired enough data to achieve statistical significance. Another advantage of the manuscript-based thesis is that you
receive feedback from the reviewers of each article during the publication process and get validation from your peers as you progress into your
PhD instead of waiting until the very last minute to ascertain novelty and
significance.

MULTIPLY AUTHORED PAPER

Institutions have rules regarding using multiply authored articles, especially when cosigned by more than one PhD student. Examples of such

rules include prohibiting a manuscript from being reproduced in more than one doctoral thesis. Another common rule is that you should not reproduce a publication for which you are not the lead author. The philosophy behind such a rule is that writing a dissertation is a critical milestone of getting a PhD: you should not take credit for someone else's writing. If you are the first of many authors, the school might require that coauthors sign off on your contribution. Make sure you prepare such forms after each manuscript is published, and gather all signatures before that brilliant postdoc decides to move to another country and away from academia. Tracking all coauthors two days before the deadline for submitting your dissertation adds unnecessary stress to the process.

HYBRID DISSERTATION

The third style is called the hybrid dissertation, somewhere between the article-based and the traditional dissertation: some chapters reproduce published papers, and some describe unpublished results or research you performed as an important-but-not-first author. Not all schools permit this hybrid style—inquire your school library about possible formats.

Befriend the Administrative Assistant   Almost anything is possible with the proper form. Want to change the title from what was suggested at the thesis proposal stage? There is a form for that. Want to change co-director? Submit your thesis in English as opposed to the school's official language? Change from a traditional to a paper-based thesis? There are forms for that. Make sure you are on top of paperwork with the help of the administrative assistant in your department. The admin should be your first call when the end becomes in sight. From the start, you should get acquainted with the graduate student administrative person to ensure you comply with deadlines on completing your classes and organizing your comprehensive exam.

10.3   WHAT A THESIS IS NOT

In figure 2.6, we had already concluded that a PhD is not a collection of several master's projects. In the same vein, a thesis is not a collection of papers. Several papers on separate topics do not constitute a dissertation.

Instead, there must be a common thread that contributes to pushing the boundary of knowledge further. Furthermore, the common thread must be scientific, not a long, quiet river. A PhD thesis is not a diary of your journey; the document should revolve around a strong common thread supporting the thesis statement and highlighting your novel and significant contributions to knowledge.

NOT AN ENCYCLOPEDIA

Rumor has it that famous physicist Louis de Broglie's doctoral dissertation was only a few pages long, despite being worthy of a Nobel Prize. Archives from Sorbonne University (France)[4] show that the length of his PhD thesis was ≈ 100 pages. However, de Broglie's most important contribution, which consisted in linking the wavelength λ of a particle to its momentum p through Planck's constant h:

$$\lambda = \frac{h}{p}$$

appears in a half-page publication in *Nature*.[5] In half a page, de Broglie changed our world from classical to quantum. When it comes to theses, longer is not necessary better—especially from the reviewer's point of view! Some institutions even limit the number of allowed pages: find out what the limit is for your institution. In engineering, doctoral theses average between 150 and 180 pages. A much longer thesis might indicate that you need to improve your writing skills.[6] If English is not your first language, consider using a phrasebook containing examples of compact academic sentences.[7] Moreover, begin with simple, boring sentences: make it pretty after it makes sense.

A thesis is neither a compendium of factoids, a textbook, nor a who is who of the scientific community. You need not re-explain every technological advancement since the invention of the wheel—use the reference section wisely.

NOT A COAUTHORED MANUSCRIPT

A thesis is not a coauthored paper with your advisor. During your PhD, the role of your advisor is to guide you toward making the most significant and impactful research and meeting—or exceeding—the objectives stated in your thesis proposal. As you write your thesis, your advisor's role is to guide you toward a document that does justice to your results. It is not, however, to impart his or her writing style to the manuscript. Furthermore, it certainly is not about proofreading it for you. When I review an article for publication, I change every word until satisfied that it meets my lab's writing standards and perhaps also reflects my writing style. However, I will not rewrite a student's doctoral (or master's) thesis as it is their document to write, with their sole name on it. Of course, I will highlight sentences that are unclear, too long, or missing an important point, and often—and uncontrollably so—provide suggestions for clarity, indicate missing information, improve the flow, and fix typos here and there. Nevertheless, I will not *entirely* change the style of the manuscript.

Interestingly, you need not obtain your advisor's approval to submit your thesis. However, securing consent helps as this guarantees at least one ally on the jury. Indeed, you could, in theory, organize a thesis defense without the support of your advisor. However, I would argue against this in the strongest possible terms. For a jury to grant a PhD, the vote must be unanimous. You would need extraordinary circumstances to change your advisor's position within the short duration of a thesis defense. Whatever disagreement you have before submitting your thesis is best solved through either of these approaches:

science: e.g., holding a thesis committee meeting to gather the opinion
    of other senior scientists in the field;
diplomacy: e.g., with the help of the head of your department; and
mediation: e.g., with the help of your school's ombudsperson.

Ideally, in the presented order: try not to kick down the ombudsperson's door at your first hiccup, but know that this mediator works for you and can provide you with tools to smooth out difficult situations.

NOT A SPRINT

Writing a dissertation is not a sprint but a marathon that some would even call the 42,195-meter hurdles. As soon as your thesis proposal is accepted, you should start thinking about your dissertation. Gradually build a slide deck with important figures and results. In addition to documenting your work, having a deck ready allows you to give presentations at a moment's notice and increase the visibility of your work. Very early on, consider downloading and familiarizing yourself with your school's thesis template. If you plan to write a paper-based thesis, gradually incorporate papers into the template as they are published.

You only control some aspects of the marathon, not all of them. If you—and you should—seek comments from your advisor, discuss how they want to proceed. Some are fine with reading chapters one at a time. Some require the whole document at once. Most prefer hearing about your outline before they are handed a two-hundred-page magnum opus. They will need some time to return your thesis for feedback: ask for a guesstimate on their part. Once you receive feedback, iterate with your advisor until you have a document you are proud of, and then submit it to your department for distribution to the jury. Most departments require some time for processing and guarantee committee members a minimum of four to six weeks to report on the document. Committee members will then send their reports to the thesis committee chairperson, who, upon reading your thesis and reports from other committee members, will decide whether:

- the thesis may be defended—the jury may ask for minor modifications before the defense; some schools require that the document be modified before the defense;
- the thesis should be returned for major corrections, after which the thesis will be reevaluated. Take this warning seriously, as you typically are granted only one second chance at submitting your thesis; or
- the thesis is rejected—the candidacy ends at this point.

Once the chairperson agrees to hold a thesis defense, it must be announced to allow other interested scientists to attend. Most departments require at least a week's notice. This last delay aims to ensure that your work's novelty and significance are validated by the most knowledgeable minds in your field.

While every mid-career faculty I know has heard of one case of failure at this stage, a thesis is seldom rejected, especially when submitted with your advisor's assent and some work has been published along the way. Therefore, booking a thesis defense date, and possibly a conference room, at the time of your thesis submission is usual enough not to jinx the results. A rejected thesis causes such a stir: canceling a booking will not be noticed in comparison. However, be careful when booking your parents' plane tickets for the defense or a vacation right after: the examiners must take as long as necessary to evaluate the science and should not be bothered with other considerations.

SOMETIMES A SPRINT

In no other section of this book is it more important to do as I suggest rather than as I did. My story reveals that, in some extraordinary circumstances, it is possible to wrap things up relatively rapidly. For instance, if the job of your dreams depended on your speedy graduation or if you were dangerously approaching your baby's due date, some protagonists may go out of their way to turn comments and evaluations rapidly. You should, however, not count on it, and incompressible delays can (almost) never be squeezed—it goes without saying. However, if you find yourself in such shoes, let people know. Perhaps years of being a good lab and department citizen might play in your favor, or perhaps, like me, you are simply surrounded by tremendously generous individuals.

10.4   DEFENSE OR *VIVA VOCE*

The thesis defense is the last occasion for the candidate to present findings to a jury to discuss the results presented in the dissertation and verify the candidate's mastery of the field. In the United Kingdom (UK) and some other countries, this oral examination is called the *viva voce*, Latin for live voice, or simply *viva*. In some countries, including the US and Canada, the thesis defense is held publicly and must be advertised. In others, such as the UK, the *viva* is held behind closed doors. In Finland, instead of a jury, you face a single opponent.[8] The thesis defense is typically a solo exercise.[9]

JURY

As you progress from your thesis proposal to your defense, the composition of your thesis committee might fluctuate to reflect the new research avenues you are charting. As you get closer to your thesis defense, your advisor will make a recommendation regarding the final composition of your jury. As usual, regular and fine prints in your school's rule book shall prevail. However, a doctoral jury typically comprises two examiners (also called opponents), a presider (also called chairperson), and a representative of the dean, that is, someone to verify that rules are being followed. More precisely, the jury consists of

an external examiner: out of the two opponents, one comes with specific expertise in your field or research from another institution or even a different country. The external examiner provides a report on your thesis, attends the defense to ask questions, suggests modifications, if applicable, and participates in the verdict;

an internal (departmental) examiner: the other opponent comes with general expertise in your field of research and may come from your department, but not from your research group. The internal examiner also reviews the dissertation, produces a report, attends the dissertation, asks questions, suggests modifications, if applicable, and participates in the verdict;

a chairperson: a third examiner, usually from the candidate's department, acts as the chairperson for the committee, having access to the thesis, producing a report, and making a recommendation based on all reports as to whether or not the dissertation is of sufficient quality to warrant a thesis defense. During the defense or *viva*, the chairperson explains the process, moderates the discussion, asks questions, and participates in the verdict;

a delegate of the head of the department or dean: the final committee member is a faculty from the school—knowledgeable on all aspects of a thesis defense—verifies that rules are correctly followed. This person has access to your thesis, is present at the defense, and may ask questions, but does not write a scientific report. The representative of the dean, however, must attest to compliance with all rules and procedures; and

the advisor and co-advisor: in some countries, such as the US and Canada, the advisor and co-advisor are part of the jury. They have reviewed your thesis before submission, and, as such, they do not typically provide a written evaluation to the thesis committee's chairperson. They are present in the oral examination and are encouraged to ask questions, albeit after the opponents and chairperson have asked their questions within each round. The thesis advisor and co-advisor count as one person, that is the co-advisor can not substitute for another jury member. In the UK, the research supervisor and co-supervisor do not actively participate in the *viva*. They act as observers only, as the *viva* is usually held behind closed doors.

The jury's composition must be approved *a priori* to ensure no conflict of interest exists between members, and all examiners are at arm's length with the candidate, advisors, and project. The concept of academic independence (see definition 2.9) usually applies. Make sure you know who the members of your jury are. Perhaps, read a paper or two from their respective research groups to familiarize yourself with their research and interests.

PREPARATION

The weeks between submitting your dissertation and defending it are just enough time to prepare your thesis defense (and finish some papers, draft some patents, finish grading papers for the class you are TA-ing, put the last touch on this new postdoctoral fellowship, and so on). Note that for a manuscript-based thesis, you are often required to have submitted manuscript(s) prior to the dissertation. Do not underestimate the time required to prepare for your thesis defense or *viva*. Review the notes you took while attending defenses in your department or discuss with students who just successfully completed their PhDs. If your doctoral examination includes a short lecture, discuss the outline of your presentation with your thesis advisor, prepare your slides, and practice with lab members. Ask them to grill you with questions. In the ideal scenario, your lab mates ask the most challenging questions so you can adequately prepare and not be surprised at the actual main event. In any case, try to anticipate questions you may be asked and, perhaps, prepare some supplementary materials for the elephant-in-the-room type of question. Include slides for material pruned out of the main presentation, just in case. Perhaps, prepare an index of all these slides with hyperlinks to avoid anxiously scrolling through many appendices to answer an examiner.

Lock in the slides a few days before the defense to allow yourself to rest before the ultimate event of your student career. Then, be ready to discuss the following points:[10]

- What were the main limitations of your field of research before your contributions?
- What is your thesis: question, hypothesis, objectives, and assumptions?
- What are your most significant and original contributions?
- What was your methodology, and why is such methodology best suited for your work?
- Where does your thesis fit within a broader context (i.e., the Big Picture)?
- What are the short-, mid-, and long-term implications of your thesis?
- What are the strengths and weaknesses of your study? Being able to discuss weaknesses shows your maturity as a critical thinker.

As you prepare for your defense, the jury evaluates your dissertation and writes a report based on whether the decision is taken to hold or not a thesis defense.

Remark 10.1: Thesis Defense   *Congrats on being invited to defend your thesis. Book a room with the suitable capacity and audio-visual (A/V) equipment for your needs. Befriend another administrative assistant to get access to the room beforehand so you can squeeze in an extra rehearsal and familiarize yourself with the layout and amenities. In addition, plan to bring the adequate dongle and a laser pointer.*

TIMELINE

Before the examination, the chairperson reviews your thesis and the examiner's reports and proceeds, or not, with the advertisement for the thesis defense. Each institution follows a specific ritual for thesis defenses. However, some aspects are common:

preparation: before the jury arrives, prepare the ground by connecting your laptop to the A/V unit to avoid the situation described in figure 10.2. Indeed, in his comic strip *PhD Comics*, Jorge Cham proposes that the number of professors needed to fix issues related to a projector is

10.2 In some thesis defenses, candidates must prepare a short lecture, requiring that they fit the past several years of their lives into a thirty-minute talk. Less without the right dongle.

$n + 1$, where n is the number of faculty members presents, and 1 is the one audience member who thinks of calling the technician.[11]

introduction: the chairperson, or president, of the thesis committee presents the jury to the audience, then presents the PhD candidate using a short bio and explains the format of the event;

presentation: the candidate proceeds to the presentation of a condensed version of their thesis, which typically includes visual support (see figure 10.2)—verify how much time you have beforehand, and curate and practice until you can comfortably deliver your lecture within the time limit;

public question period: the jury asks questions both on the presentation and on the dissertation. Members ask several questions, typically in rounds, beginning with the external examiner and finishing with the closest advisor. The president then invites the audience to join the discussion;

private question period: the chairperson asks the audience to leave the room for a private question period between the candidate and the jury;

deliberation: when questions, committee members, or the candidate are exhausted, the president asks the candidate to leave the room for the thesis committee to agree on a verdict; and results: the candidate and the audience are asked back in the room to receive the verdict.

Verdict options include:

accepting the thesis without modification—a rare bird, indeed!;

with minor modifications—typos, grammatical errors, adding some references, tweaking paragraphs here and there, ...;

with major modifications—requiring extra experiments or data analysis and, possibly, a new defense; and

rejecting the thesis—ending the candidacy, an even rarer event than the perfect thesis.

Box 10.2

Shrinking Years Down to Minutes

If your doctoral defense includes a lecture, find out how long your presentation should last and practice your defense to avoid running overtime. Rehearse your thesis defense in front of lab members and ask them to ask you tough questions. Rest before your defense, or try to: cook muffins or go for a jog or a walk. Refresh your mind. Depending on your topic and the energy level of your thesis committee, questioning may last from one to several hours. You may want to read some of your opponent's most important papers to anticipate where their questions may take you. Bring an annotated copy of your thesis, water, and comfortable shoes.

PUBLIC VS IN-CAMERA DEFENSES

In countries where thesis defenses are public, you may still obtain permission to hold it behind closed doors, *in camera*. This exceptional circumstance requires written authorization and is given when it is required to allow adequate protection of IP, typically when the research was performed in collaboration with an industrial partner. The library may also grant a slight delay before making the dissertation public.

ON ATTENDING SOMEONE ELSE'S DEFENSE

When attending a thesis defense, there are appropriate moments when you can leave the room. An astute chairperson will let the audience know when such moments arise: for example, immediately after the lecture, before the questions start, or in between question rounds. Do not leave, walk back in, then walk out again outside of these prescribed moments. If you commit to attending a defense, plan on being there in solidarity until the end: the process is sufficiently nerve-wracking for the candidate, and distractions from people walking about the room are not welcome.

ENJOY!

At no other point in your academic life will you know a subject as much as you do when you defend your PhD thesis, and very rarely will you be given such a genuinely interested audience. Given adequate preparation, you may even enjoy the moment. Do celebrate the positive outcome with family, friends, collaborators, and lab mates, albeit afterward. Do not sacrifice preparation time to plan for the after-party. Let someone else plan it for you, and do not think about it until *after* your thesis defense. Until that day, organize someone else's party, attend as many defenses as possible, or discuss with every lab member who survived their *viva*.

10.5 BEFORE LEAVING

In the very final days of your PhD, you will make the changes to your dissertation as required by the jury. If they asked for revisions that are more important than typos, you need to have such changes approved. Your school library will provide you with a checklist concerning the final formatting of your thesis. Your dissertation is finished, but the research continues. Turn in your lab books to your advisor, agree on a platform to deposit data and code, agree on a *modus operandi* for unpublished or pending papers, and perhaps, train the next student on your experimental setup. Tour the lab with your advisor or research associate to identify what can be safely discarded and what needs to be retained (for example, if peer reviewers for your recently submitted papers have additional questions). You will save the team many headaches by presenting a clear plan.

Now go print those business cards with the letters PhD on them—you deserve them!

10.3   The missing link between R&D and new knowledge and innovation (i.e., societal impact).

10.6   INTELLECTUAL PROPERTY

Aside from fame and money (not!), one key motivation for graduate students should be societal impact through advancing scientific knowledge and technological innovation.

Box 10.3
Missing Link

Figure 10.3 illustrates the progression between new knowledge issued from research and societal impact. What are the missing links? This section discusses the chain of technology maturation from lab to market.

A common misconception from researchers is that new ideas are the most crucial step in technological advancement. As a young researcher, I shared their point of view. Phrased differently, I would ask myself, how come most royalty deals involve such low percentages back to patent inventors and assignees? At the time, I was utterly oblivious to the amount of work required to advance new knowledge from the proof-of-principle phase to commercialization and, finally, societal impact. Once an idea leaves the lab, it is typically at TRL 1-3. Indeed, six more levels are required before a new technology begins to change the world.

Moving technology from level 3 to level 9 occurs under resources' scarcity as, at this stage, the work is neither fundamental nor quite ready for market yet. This phase, shown in figure 10.4, is often called the Valley of Death,[12] where many ideas die before reaching maturity.[13]

Patents and protection of IP offer an incentive for private stakeholders to fill the funding gap and invest in maturing the technology until it reaches its final stage of development. The incentive for such an

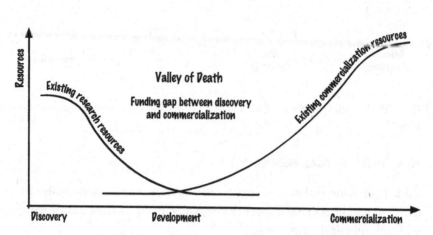

10.4 The Valley of Death between basic research and commercialization.

investment is the rights to exclusive exploitation of the technology for several years.

Disclaimer

Patent law is complex, and for fear of making a mistake, very few authors are willing to describe it in writing. As a result, students are confused about patents and IP. In this section, I describe in very broad terms the idea behind a patent, but the reader is warned that:

- patent laws vary by country;
- inventor compensations depend on institutional rules; and
- licensing agreements are subject to negotiations between the patent owner (i.e., licensor) and the entity interested in exploiting the technology (i.e., licensee) on a case-by-case basis.

This section should serve as a lexicon for discussing with your advisor and the research office at your institution.

WHAT CONSTITUTES INTELLECTUAL PROPERTY?

IP refers to the creations of the mind.[14] Its two branches are:

copyright: related to artistic creations (scientific papers, books, music, paintings, films, ...) as well as algorithms and databases; and industrial property: related to patents for inventions, industrial designs, integrated circuits, trademarks, and other signs conveying information.

Simply stated, copyright indicates that copying an original work can only be done by its author or with his or her permission. Patents and other industrial property work similarly to exclude people other than its assignee from making, using, or selling an invention in exchange for publicly describing it.

Definition 10.1: Invention   *An invention is a creation of the mind that is novel, useful, and non-obvious to someone acquainted with the field. The creators of an invention are called inventors.*

Box 10.4
Handled Snifter

A balloon is a type of glass used to sip brandy and other fine liquors. Is adding a handle to such a glass (see figure 10.5) a patentable invention? According to the United States Patent and Trademark Office (USPTO), to file for a patent, the invention must be:

novel—arguably, there is no balloon with handles, so, yes, it is novel;

useful—one could argue that for hot liquors, a handle could be useful; and

non-obvious—for anyone designing stemware—or anyone, really—the obvious solution to drinking a hot beverage is to add a handle.

The design of a balloon with a handle should not be patented unless a talented patent agent argues that this particular design allows for brand recognition and files for a design patent.

Definition 10.2: Patent   *A patent is a negative right in that it allows its owner the right to exclude others from exploiting a technology. A patent must include a description of the invention that is sufficiently detailed to allow for reproduction of the work.*

Nowadays, new technological knowledge is seldom produced in one's garage, and, as a result, patents involve several stakeholders:

inventors: individuals who intellectually contributed to the creation of the invention, including students and their advisors. On the contrary,

10.5 A sketch of a handled brandy balloon snifter. Is this a patentable invention?

people who simply executed work under instructions or who were
present in the room when it happened do not qualify;

assignee: parties (individuals or entities such as universities or research
centers) who own the rights to the invention;

patent agent: a lawyer specialized in all matters related to patent law,
including writing the invention's description for patent filing;

licensee: parties (individuals or entities) who have secured the rights—
exclusively or not—to exploit an invention over some number of years,
in some countries and/or for some applications; and

licensor: parties (individuals or entities) who are providing the IP.

Box 10.5
IP Ownership

What do you know about IP ownership at your institution? Is student
entrepreneurship encouraged? Are you looking for an opportunity to join a
start-up upon graduation?

Filing for a patent is not cheap, to say the least. Even the wealthiest
universities cannot afford to patent every novel, useful, and nontrivial

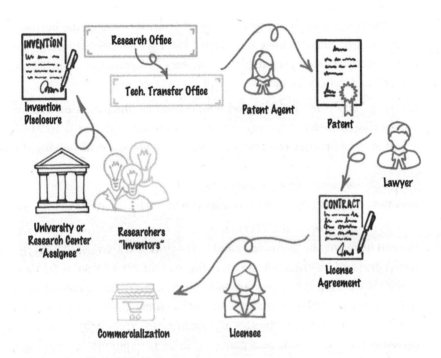

10.6 The patent cycle from invention to commercialization.

idea their researchers invent. So, instead, they rely on a TTO to decide which technology has the greatest chance of commercial success based on market studies and their expertise and network of potential licensees. Figure 10.6 shows a typical pathway from an idea to commercialization through the various offices of a university. First, researchers disclose their inventions to their university's research office through a document called invention disclosure. Next, the research office acts as a liaison with the TTO, which decides what protection is the most adequate given their market knowledge. Finally, if the TTO decides to go ahead and protect the invention as a patent, someone called a patent agent converts the invention disclosure into a patent application and files it to one or several patent offices, according to the number of countries targeted for protection. While the patent is being evaluated, the patent is said to be pending or awaiting a decision.

Typically, a TTO pays for all fees upfront, so it tries to rapidly find a commercial partner interested in exploiting the technology in exchange for some compensation. Such a commercial partner is called the licensee. A license agreement allows the licensee to exploit the technology for commercialization in exchange for monetary compensation. The license agreement typically includes:

lump sup: allowing the TTO to reimburse itself for the patenting fees;
commercial milestones: to ensure that the licensee exploits the technol-
    ogy as soon as they can and make sure they are not just shelving the
    patent;
a stake in the company: such as shares; and
royalties: a percentage of the revenues generated by the new technology.

This money flows to the TTO to reimburse the cost incurred for protecting and licensing the technology and to the university to reward both the institution and inventors. As an inventor, you may receive some fraction of the royalties after expenses are paid, but very few receive a check before they finish their PhDs. Indeed, if you add the time required to file for the invention disclosure, file the patent, find a commercial partner, reach an agreement, let the commercial partner gear up to exploit the technology, market it, sell it, gain revenue from it, report to the TTO (typically at the end of their fiscal year), let the TTO to report to the school (at the end of their fiscal year), and the school to cut a check to you (at the end of their fiscal year), you have plenty of time to finish and defend your dissertation. You might even be the commercial partner, exploit your entrepreneurial fiber, and launch a company based on your invention, as very few people know it more than you do. However, I suggest you do not retire just yet. Most students' most crucial asset gained during a doctoral project is their PhD degree. However, on the off chance that you are the exception—that your invention will revolutionize the world as we know it—familiarize yourself with the process, ask questions regarding IP protection, and be somewhat involved in its translation to the industry.

III

TOOLS OF THE TRADE

11

WRITING TIPS

"Publish without perishing."

Peter J. Feibelman[1]

Writing is one of the most important skills to master in academia. Indeed, you are reading this because you were convincing in your graduate school application essay, and perhaps also in your scholarship dossier. You will perform experiments because your advisor was successful in getting grant money for equipment. And you will travel to exotic (and not so exotic) destinations to present your conference papers. Throughout the book, I have scattered writing tips where they were the most relevant. In this chapter, I wish to delve into some outstanding, yet important, themes that did not find a home elsewhere in this book. For even more tips, consult *The Scientist's Guide* to Writing by Stephen B. Heard[2]—an important reference for all STEM PhDs.

11.1 ANATOMY OF AN ABSTRACT

Abstracts are to the scientific world what trailers are to movies. The reader decides whether or not to read the whole document after consulting the abstract. They are found at the beginning of every scientific manuscript, conference proceeding, or dissertation. Sometimes they are the only document used by program committees to decide whether or not to invite you to their conference. The purpose of the abstract is twofold:

- enticing the reader to continue to the rest of the article; and
- announcing the take-home message.

Remark 11.1: Hard Limit   *Most publishers fix a hard Limit on the word count for an abstract. Find out what that maximum is before writing, and stick to it.*

Abstracts should follow the general structure highlighted in the next section. Particularities of the thesis proposal's abstract are mentioned in a later section.

STANDARD RECIPE

Some scientists are tremendous authors. However, the rest of us must constantly develop skills to communicate our results efficiently, often in a foreign language. Until you become so proficient that you build your own style, consider this general recipe: summarize each section of your document in one or two sentences. For example, write:

context: one to two sentence(s);

scientific question or problem (or current limitations): one sentence, often beginning with "However,";

thesis: one sentence, typically beginning with "Here, we [show, demonstrate, prove,...]";

materials and methods: one sentence;

most important result: one sentence;

discussion: one sentence; and

conclusion and outlook: one sentence.

The above recipe is your starting point. Once gathered, combine the sentences to shorten your document and comply with the word count limit. Write your abstract last. Indeed, to describe a section, you must know what it entails. Composing an abstract is far from easy. Several iterations often are required to include all relevant information in a compact manner.

THESIS PROPOSAL RECIPE

The abstract recipe for a thesis proposal requires adapting its ingredients and proportions to account for the prospective nature of the exercise:

context: one to two sentence(s);

scientific question or problem (or limitations): one sentence, often beginning with "However,";

general objective: one sentence, typically beginning with "Here, we propose";

planned materials and methods: one sentence;

preliminary and anticipated results: one sentence; and

expected impact: one to two sentence(s).

CONDENSING EACH SECTION

Here is a proposed strategy to compose a fine abstract. With time, you will develop your own style. Until then, I suggest working on an example inspired by my lab's research. (In a typical academic fashion, I rarely relinquish an opportunity to talk about my work.)

Context   The context is where you situate your work. How much detail? Always start with your audience. If you write an abstract to be evaluated by your thesis committee in, say, electrical engineering, you may take it for granted that this information is known. "This work is about electrical engineering" is too broad and, therefore, is a waste of words. However, if you applied for a grant in a general science and engineering council, situating the reader might be a good idea. "Electrical engineering has contributed to improving medical imaging through..." is a proper first sentence.

The first sentence should describe a concept familiar to your readers as a springboard to an exciting place. If it is not the case, add another sentence to lead to that concept. The first sentence is so important that it rarely is written at once (I, for one, am not talented enough to achieve this!). Begin with a series of simple sentences, that is, a subject and a predicate:

Optical fibers allow performing optical microscopy in vivo.

subject   *predicate*

What else would be considered background? Add another simple sentence: "Optical fibers were originally designed for telecommunications." Finally, open the topic with yet another simple sentence: "Optical fibers enable endoscopy." Writing these three sentences one after another would be grammatically correct but perhaps a little ordinary. After settling on the main ideas, merge the sentences into: "First designed for telecommunications, optical fibers have allowed optical microscopy techniques to be performed in vivo through devices called endoscopes." This acts as the abstract's opening line.

Scientific Question (Limitations) The knowledge gap or technological limitation is akin to the disruptive event in popular literature. The Count of Monte Cristo is put in jail. Hagrid shows up on Harry's doorstep. Here, current optical fibers are not adequate for endoscopy. Something is either not well understood or the previous iteration failed. With the scientific question or limitation statement, you are trying to answer the question "Why did you tackle this work?" It is not uncommon for this sentence to begin with "However."[3]

In the current example, the *disruptive event* could revolve around: "Unleashing the full potential of endoscopy requires designing optical fibers to meet the specific constraints of medical imaging." The last sentence is a somewhat soft statement. A more disruptive idea could read like this: "Unless explicitly designed for medical imaging, optical fibers will fail at unleashing the full potential of endoscopy." An even bolder statement would read as follows: "Existing optical fibers are not good enough for endoscopy," but this risks alienating people who published on the topic before (read: senior scientists who are very likely to review said abstract). Consider the scientific question (or limitation) as a disruptive event but without the drama associated with popular literature.

Thesis or General Objective One of the most important sentences of your abstract answers the question "What are you presenting?" by summarizing the thesis of your paper or the general objective of your proposal. Introduce such a sentence by using words that deliberately break the flow. In the body of the manuscript, one could write "In this paper, we demonstrate (or show, or introduce ...)." For an abstract, use the short version "Here, we present" as it provides a more compact formulation. The general objective of the fiber example could read like: "Here, we present a novel double-clad fiber coupler explicitly designed for the constraints of endoscopy." For your thesis proposal, you must prospectively announce your objective: "Here, we propose to develop a new fiber optic coupler to demonstrate its impact on the resolution and contrast of a single-fiber endoscope."

Materials and Methods   The abstract summarizes several months of experiments in one sentence to answer the critical question: *How did you do it?* Only use the key points. Following our example: "The optical fiber coupler was built through the fusion and tapering of commercially-available double-clad fibers." Here, techniques are mentioned, but details are relegated to the core of the manuscript. For example, the abstract is exempt from fiber diameters, types of glass, fusion temperatures, and so on. In the context of your research proposal, mention the essential materials and methods you plan on using without the nitty-gritty of your experiment. For example, if you plan to use a laser etching technique, writing "a pulsed laser etched a waveguide into a silica substrate" is plenty detailed. The much heavier version: "a 150 femtosecond laser (Brand from Manufacturer (City, Country)) will be used to etch a 155x228 micron waveguide into a custom-polished borosilicate glass (BK7) prism from Other Manufacturer" is too detailed for a two-hundred-word abstract, but belongs in the full proposal or article.

Results   The result sentence answers the question: "Did you achieve what you announced two sentences ago?" For the current example, we get: "The novel assembly improved the signal-to-noise ratio by a factor of 100, did not affect the lateral resolution, but degraded the axial resolution by 15%."

For a thesis proposal, you must describe both preliminary and anticipated results. For example, "Simulations showed that the new optical fiber should improve the signal-to-noise ratio by a factor of 120 and should only degrade the resolution by 20%."

Conclusion, Outlook, and Impact   You conclude with the extent to which objectives were reached. You may also point out the direct impact on the field and society. For example, "The signal-to-noise ratio enhancement demonstrated here paves the way for diagnosis of smaller lesions at a stage where the prognosis is improved."

11.2   LEARNING BY DISSECTING

As you read papers, take the habit of dissecting abstracts into their components. For example, Figure 11.1 shows a short five-sentence abstract,[4] each summarizing a section of the publication it announces.

Box 11.1
Dissection

As you are reading papers from other authors, identify sections of an abstract. Do you notice how often the phrases "however" and "here we" are used?

Remark 11.2: *(Copy Editing) The abstract reproduced in figure 11.1 used the acronym green fluorescent protein (GFP) without defining it. It should have been. Despite being reputable, not all journals provide copy editing and,*

Multiplexed two-photon microscopy of dynamic biological samples with shaped broadband pulses

Rajesh S. Pillai,[1] Caroline Boudoux,[1,2] Guillaume Labroille,[1]
Nicolas Olivier,[1] Israel Veilleux,[1] Emmanuel Farge,[3]
Manuel Joffre,[1] and Emmanuel Beaurepaire[1]

[1] *Laboratoire d'Optique et Biosciences, Ecole Polytechnique, CNRS, and INSERM U696,*
91128 Palaiseau, France,
[2] *Ecole Polytechnique, Montreal, Canada,*
[3] *Institut Curie, CNRS, 75005 Paris, France*
manuel.joffre@polytechnique.edu, emmanuel.beaurepaire@polytechnique.edu

Abstract: Coherent control can be used to selectively enhance or cancel concurrent multiphoton processes, and has been suggested as a means to achieve nonlinear microscopy of multiple signals. Here we report multiplexed two-photon imaging *in vivo* with fast pixel rates and micrometer resolution. We control broadband laser pulses with a shaping scheme combining diffraction on an optically-addressed spatial light modulator and a scanning mirror allowing to switch between programmable shapes at kiloHertz rates. Using coherent control of the two-photon excited fluorescence, it was possible to perform selective microscopy of GFP and endogenous fluorescence in developing *Drosophila* embryos. This study establishes that broadband pulse shaping is a viable means for achieving multiplexed nonlinear imaging of biological tissues.

— Context
— Thesis
— Methods
— Results
— Conclusion

11.1   A dissected abstract. Adapted from Pillai et al. and reproduced with permission from Optica.

too often, mistakenly rely on authors, reviewers, and editors to catch errors and typos.

STRUCTURED ABSTRACT

Some journals ask for structured abstracts[5] using headings to announce the abstract's content explicitly. In a structured form, the abstract shown in figure 11.1 might look like this:

Introduction: Coherent control can be used to enhance or cancel concurrent multiphoton processes selectively, and has been suggested as a means to achieve nonlinear microscopy of multiple signals.

Aim: Here, we report multiplexed two-photon imaging *in vivo* with fast pixel rates and micrometer resolution.

Methods: We control broadband laser pulses with a shaping scheme combining diffraction on an optically-addressed spatial light modulator and a scanning mirror, allowing switching between programmable shapes at kilohertz rates.

Results: Using coherent control of the two-photon excited fluorescence, it was possible to perform selective microscopy of the green fluorescent protein and endogenous fluorescence in developing *drosophila* embryos.

Conclusion: This study establishes that broadband pulse shaping is a viable means for achieving multiplexed nonlinear imaging of biological tissues.

The strategy exposed here is only a starting point. Everyone eventually develops their own writing style and method.

11.3 ACRONYMS

Are acronyms permitted within an abstract? Some journals allow for them, while others do not: make sure you review the journal's guidelines. In the absence of specific policies, here are some general rules. If an acronym is used in the abstract and the document's body, it must be defined where it is first used in the abstract and first used in the body. Only define the acronym in the abstract if you use it more than once, which is unlikely if the word limit is tiny.

While we are at it, let us review some rules about acronyms. You should avoid using acronyms in the title of any document to ensure that your paper is accurately indexed. In the body of the text, a rule of thumb is to define an acronym only when it is used at least three times.[6] Indeed, it is easier to read a three-word phrase than to remember its acronym. However, one should not wait for an abbreviation to have been used three times before defining it!

Box 11.2
Definition of an Acronym

Acronyms must be defined where they are first used. They are specified within parentheses following their first mention or within square brackets when the definition occurs within parentheses, as in: "Imaging techniques such as magnetic resonance imaging (MRI) and computed tomography (CT) have revolutionized medicine. Both use advanced reconstruction techniques (such as a fast Fourier transform [FFT])."

For large documents, one should provide a table of acronyms for easy reference. Some scientific documentation applications, such as LaTeX, include packages tracking acronyms to ensure that they are defined on the first occurrence. Such packages may also assist in creating a table of acronyms. You must define every abbreviation, even if it is uber well known in your field, as you never know when someone from another area might seek inspiration from your paper. The only exception is for acronyms that have become words. Examples of such acronyms are IQ, HIV, and FAQ. In other terms, unless your acronym counts for Scrabble, you must define it. Finally, the article used before your acronym depends on its pronunciation: if it sounds like a vowel, use "an," but if it sounds like a consonant, use "a."

ARTICLES AND ACRONYMS

In "An MRI system is used on a US patient," the consonant "M" is pronounced "em," which sounds like a vowel, while the vowel "U" is pronounced "you," which sounds like a consonant. Not all acronyms are pronounced letter by letter—some, like random-access memory (RAM), are pronounced as words.

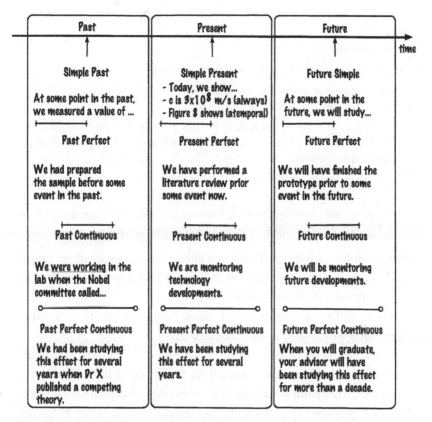

Past	Present	Future
Simple Past	**Simple Present**	**Future Simple**
At some point in the past, we measured a value of ...	- Today, we show... - c is 3×10^8 m/s (always) - Figure 8 shows (atemporal)	At some point in the future, we will study...
Past Perfect	**Present Perfect**	**Future Perfect**
We had prepared the sample before some event in the past.	We have performed a literature review prior some event now.	We will have finished the prototype prior to some event in the future.
Past Continuous	**Present Continuous**	**Future Continuous**
We <u>were working</u> in the lab when the Nobel committee called...	We are monitoring technology developments.	We will be monitoring future developments.
Past Perfect Continuous	**Present Perfect Continuous**	**Future Perfect Continuous**
We had been studying this effect for several years when Dr X published a competing theory.	We have been studying this effect for several years.	When you will graduate, your advisor will have been studying this effect for more than a decade.

11.2 Verb tenses applied to academic writing.

11.4 VERB TENSES

Figure 11.2 reproduces the diagram I have drawn the most on my office white board.[7] As in other forms of writing, present tenses describe current events, and past and future tenses describe past and future events, respectively. In addition, use:[8]

past tenses: for citing literature, past events, and measurements made;

present tenses: for citing current frameworks and events, general truths, references to figures and tables, and other atemporal facts, such as constants; and

future tenses: for proposed research, future avenues, and research perspectives.

Box 11.3

Past vs Present

The following sentences employ different verb tenses:

- The temperature varied linearly with time.
- The temperature varies linearly with time.

The first case uses the simple past to describe a single data point, while the second case expresses a general truth, implying that the temperature *always* varies linearly with time.

A thesis proposal contains all such verb tenses. Here are common uses of verb tenses (indicated within [ ]) for sections of a proposal:

literature review: confocal microscopy allows [simple present] for optical sectioning. In 2005, Boudoux et al.[9] showed [simple past] a video-rate spectrally encoded confocal microscope;

statement: dedicated optical fibers allow [simple present] endoscopes with sufficient SNR to detect small lesions;

methodology: we modeled [simple past] the behavior of light using Eq. 3 in which c stands [simple present] for the speed of light. The instrument acquired [simple past] data at video rate. Before data acquisition, we had prepared [past perfect] the sample with a solvent. For the proposed research, we will use [future simple] this technique;

results: figure 2 shows [simple present] data along with a regression curve. For the proposed work, we anticipate the results will show [future simple] a linear trend; and

conclusion and outlook: This research paves [simple present] the way for new adventures (statement). In future work, we will attempt [simple future] to improve the resolution.

Active vs passive voices   Describing an action is performed using one of two voices: active or passive. For example:

active: Donna Strickland won the 2018 Nobel Prize in Physics; and[10]
passive: The 2018 Nobel Prize in Physics was won by Donna Strickland;

In the active voice, Prof. Strickland is the subject of the sentence, while, in the second one, the Nobel Prize is. The passive voice is harder to read, but it is often used in science as the subject—the researcher performing the experiment—is not as interesting as the object, that is, the measurement. Too often, "we measured X" becomes "X was measured." Another more exciting way to rewrite the sentence is "the measured value of X falls within the prediction of the model." The last sentence is active and conveys more information while ignoring the researcher who performed the experiment.

11.5   DIVIDE AND RULE

Very few single-author documents are as intimidating as a doctoral dissertation, especially if you write it all *de novo*. To be digested, the two-hundred-page mammoth must be broken down into simpler bits. In their book *How to Get a Ph.D.*, Phillips and Pugh[11] suggest that there exist two types of writers:

planners or sequentialists: start from an outline and break each chapter into sections, each section into subsections, and each subsection into paragraphs before writing. Their thesis is written one paragraph at a time; and

get it all out-ers or holists: must first write to get a sense of what they want to say and produce complete drafts that are refined through several iterations.

I certainly fall into the first category: I will not tackle a chapter unless it is broken down into fifteen minute writing segments (that can easily fit in between meetings, as a researcher's life seldom allows for long writing periods). I will further fight white-page syndrome by adding dummy text and dummy figures to my elaborate outline to get a sense of each chapter's length. I use LaTeX's command

```
\textcolor{gray}{\lipsum[1-1]}
```

to compile paragraphs of *Lorem Ipsum* and use color to make sure that no dummy text is left behind. My very early draft is composed of paragraphs like this one:

Lorem ipsum dolor sit amet, consectetuer adipiscing elit. Ut purus elit, vestibulum ut, placerat ac, adipiscing vitae, felis. Curabitur dictum gravida mauris. Nam arcu libero, nonummy eget, consectetuer id, vulputate a,

magna. Donec vehicula augue eu neque. Pellentesque habitant morbi
tristique senectus et netus et malesuada fames ac turpis egestas. Mauris
ut leo. Cras viverra metus rhoncus sem. Nulla et lectus vestibulum urna
fringilla ultrices. Phasellus eu tellus sit amet tortor gravida placerat. Integer
sapien est, iaculis in, pretium quis, viverra ac, nunc. Praesent eget sem vel
leo ultrices bibendum. Aenean faucibus. Morbi dolor nulla, malesuada eu,
pulvinar at, mollis ac, nulla. Curabitur auctor semper nulla. Donec varius
orci eget risus. Duis nibh mi, congue eu, accumsan eleifend, sagittis quis,
diam. Duis eget orci sit amet orci dignissim rutrum

As writing progresses, the dummy paragraphs are replaced with original text, but so long as I began with a detailed outline, I rarely faced
writer's block.

Try following the writing sequence proposed in chapter 4. Begin with
easy sections. If no word comes to mind, sketch some figures: your experimental setup or results. Then, describe what you see in each figure. If it
is a graph, start with the obvious: "Fig. 1 plots Y against X." If you still do
not know what to write, describe the figure aloud and record your speech.
If you still lack inspiration, read more. Finally, when English is not your
first language, open a phrasebook for scientists.[12]

Most writers, including the prolific Stephen King, Leo Tolstoy, and
John Steinbeck, will argue for writing every day. One way to follow their
advice is to write a two-line summary for every article you read.

11.6   THE ROYAL I

The most important pronoun of your scientific career is "we." In fact,
despite having authored several books solo, I often describe the content
using the royal we: "we wrote, we discuss, we say" from sheer habit, and
perhaps also to acknowledge the extensive list of collaborators. However,
during a thesis defense (or a faculty position interview), it is important the
committee understands what is your individual contribution, especially if
you operate in large teams publishing several-author papers. For collective
efforts, and to avoid looking like a *prima dona*, use "we," but resort to using
"I" occasionally to highlight important personal contributions.

12

ROADBUMPS AND BARRIERS

"We choose to go to the moon in this decade and do the other things, not because they are easy, but because they are hard."

John F. Kennedy

"Completing a PhD is hard, but it does not have to be that hard."

Caroline Boudoux

On the first week of my doctoral studies, I was advised to take a walk and watch the sunset on the Charles River on days when the stress would become unbearable—an exam, a deadline, a rejected paper—on the promise that, later, I would remember the sunset and not the stressful situation. I did—several times! And twenty years later, I confirm that I remember several colorful skies and have (mostly) forgotten the reasons behind the walks. A PhD is a tough endeavor. Please do not listen to anyone who claims to the contrary—they probably only remember sunsets. However, the mere fact that I remember so many is a testament to how difficult the journey has been. Pushing at the limit of knowledge is hard, but it should not be insurmountable. Help exists—this chapter aims to openly address common barriers and challenges encountered by students in their academic journey in the hope of de-stigmatizing these hurdles and suggesting avenues for support.

12.1 REPRESENTATION MATTERS

In the mid-1960s, David Wade Chambers designed the draw-a-scientist test (DAST) to investigate children's perceptions of what a scientist looks like.[1] For generations, school-age boys and girls have drawn a white male with messy hair, glasses, and a lab coat. Indeed, out of nearly five thousand drawings in the original study, only twenty-eight portrayed women. Why does this matter? Because when people do not see themselves represented in science-related activities and careers, it becomes harder to envision their future selves as scientists and for others to believe in their potential.

This simple experiment shows that some groups of people are missing from the collective imagination of who belongs in STEM. In most Western countries, the diversity of engineering programs and faculties is not representative of the diversity of the general population. In Montréal, for instance, in 2023, women account for (slightly) more than half of the general population but make up only 30 percent of PolyMtl undergraduate students and less than 20 percent of faculty members.

BIPOC, persons with disabilities, and lesbian, gay, bisexual, transgender, queer or questioning, and two-spirit (LGBTQ2S+) folks are also significantly underrepresented in STEM. Together with women, some of these groups have been designated by Canada's Employment Equity Act as facing additional barriers in the workplace. Indeed, when climbing the academic ladder:

- BIPOC[2] and women scientists[3] receive, on average, less research funding;
- female researchers are, on average, less cited than male researchers, even when publishing in top-tier journals;[4]
- LGBTQ2S+ professionals in STEM are 30 percent more likely to experience workplace harassment compared to their non-LGBTQ2S+ peers;[5] and
- as alluded to in figure 5.6, letters of reference use different lexical fields (females are *dedicated* and *hardworking*, while males are *excellent*

and *unique*), are shorter for females, and include doubt raisers (for women) instead of discussing accomplishments such as patents and publications (for men).[6]

If you are part of an underrepresented or equity-seeking group, do not despair: the wind is (slowly) turning. Indeed, women scientist representation in the DAST rose from less than 1 percent in the 1970s to a third by 2016.[7] Furthermore, funding agencies and universities are becoming more numerous to recognize that: "...to tap into the widest possible range of ideas to foster innovation and creativity, we need to engage the entire population, in all its diversity" (PolyMtl's IDEA training).

Such is the aim of the IDEA movement based on the following pillars:[8]

inclusion: a sense of well-being and belonging that promotes authenticity and fulfillment;

diversity: identity factors, visible or invisible, that characterize an individual or group;

equity: measures to guarantee equal opportunities by ensuring fair treatment; and

accessibility: the conditions that allow any individual, regardless of their cognitive or physical abilities, to access and enjoy a place, product, or service and to do so autonomously.

To build a more inclusive, diverse, equitable, and accessible academia, we must first acknowledge (and then deconstruct) the unconscious biases influencing our decisions.

UNCONSCIOUS BIASES

"It is easier to disintegrate an atom than a prejudice."

Albert Einstein

Our brain is fed a torrent of information every second, much more than it can realistically process. Quick decisions are often based on mental shortcuts made from implicit associations that can influence our emotional or rational responses in everyday situations. Unconscious biases, such as anchoring,[9] are often exploited by marketing. For example, when making a purchase, we rely on the first piece of information, the anchor

12.1 Unconscious biases affect how we see others but also how we internally talk to ourselves. Be kind to yourself.

(e.g., the manufacturer's suggested retail price), as being the truth, such that we perceive any discount as a good deal, despite the real value of a product.

Unconscious biases affect every one of us. They can interfere in recruitment, evaluation, and decision-making but also during peer review, the cornerstone on which the pursuit of knowledge is founded, leading to skewed judgments and reinforced stereotypes. Examples include:

prototype-based bias: having a *model* candidate in mind (c.f. DAST) for a position and only considering people who perfectly match this idealized profile (who may or may not, in fact, exist); and

halo or horn effect: extrapolating a single positive (halo) or negative (horn) trait to judge a person without taking into account their other characteristics. For example, making assumptions about the quality or competence of a researcher solely based on their affiliation to a prestigious (or, conversely, lesser-known) school.

Unconscious biases are so pervasive that they are effectively internalized by people and positively or negatively impacted by the stereotype they convey, either creating a sense of entitlement or a lack of worthiness. For example, figure 12.1 illustrates that someone who feels like they

12.2 Impostor syndrome—that is, the inability to realistically assess your competence and skills—is prevalent among scientists (as the more we know, the more we realize we don't know), especially for those who constitute a minority and lack role models in academia.

rightfully belong will more readily contest the validity of a statement (i.e., questioning external factors), as opposed to someone who has heard over and over that they do not *have what it takes* to pursue a STEM-related career (i.e., questioning themselves).

IMPOSTOR PHENOMENON

Impostor phenomenon or syndrome[10] describes the insecurities that capable and qualified individuals experience despite their credentials, skills, or achievements. They may have a hard time owning their success (e.g., *I had help, I was lucky*), doubt their skills and accomplishments (e.g., *I don't deserve this accolade*), and feel like a fraud who could be unmasked at any moment. Everyone can fall prey to this phenomenon[11] at any point in their career, but it is especially prevalent for people who are underrepresented in any given field. The impostor phenomenon can have many negative consequences, from missed opportunities (see figure 12.2) to seriously altered professional satisfaction, which may trigger anxiety and depression.[12] In graduate school, impostor syndrome is so prevalent that you could discuss it with as few as ten friends without running the risk of being alone. Ironically, before hearing about the impostor syndrome, junior scientists may feel that such insecurities are unique. While there is no magic cure—and I can attest that it does not spontaneously resolve with age, diplomas, or accolades—cultivating a support network can help calibrate your perspective when you start doubting yourself.

The effects of the impostor phenomenon can be exacerbated by a hostile or unfriendly environment, like an unwelcoming lab culture. If you are struggling with stress or performance anxiety, consult your school's student office to seek psychological support. A mental health professional may help you pinpoint internal and external causes of feelings of self-doubt and propose strategies to mitigate them.

Box 12.1

Taming the Inner Voice

Once in my life, impostor syndrome played in my favor. When I was invited to interview for a doctoral school, I was convinced that the committee had made a mistake. I told myself I would enjoy the trip and arrived at the interview completely relaxed. In my mind, the stakes could not have been lower as this whole thing was—as the committee would surely realize—a mistake. I played the part and even politely joked with the panel, which, in retrospect, must have made some impression. To my utter surprise, the invitation was not a mistake and I was indeed accepted into the program. This one time, we, the inner voice and I, successfully teamed up. Most of the time, however, it is a pain to deal with: it wakes me up at night, plays every conversation or lecture back to me, and overanalyzes every new opportunity. On a more positive note, in his book *Think Again*, organizational psychologist Adam Grant argues that impostor syndrome may lead to confident humility.[13] As our competence grows, the impostor syndrome does not disappear (alas!) but instills the required amount of introspection to prevent blinding overconfidence or complacently resting on your laurels. When my impostor syndrome is paralyzing, I turn to a network of well-meaning friends and mentors whose judgment I have learned to trust and with whom I feel safe discussing ambitions and fears.

FOREWARNED IS FOREARMED

"Not everything that is faced can be changed, but nothing can be changed until it is faced."

attributed to James Baldwin

In theory, unconscious biases lose much of their power once we increase our awareness and pause to think about their root causes and effects. However, in practice, changing organizational cultures, institutionalized traditions, and individual mentalities takes time. The future looks brighter than the past, but it is up to each and every one of us to drive positive change and help create safer, fairer, and more inclusive academic spaces for all.

12.2   MENTAL HEALTH

There exists a mental health crisis in graduate education.[14] A large study[15] of 3,600 graduate students across ten campuses in California found that 30 percent of master's students and 39 percent of PhD students experienced "depressive symptoms that are severe enough to warrant a further assessment for depressive disorder." These results echo studies in other parts of the globe.[16]

> **Disclaimer**
>
> This section describes a mental health crisis in graduate education at large, and affecting engineering graduate students in particular. *This section cannot and does not constitute medical advice in any shape or form. The information presented here does not substitute for the knowledge, skill, and judgment of qualified health care professionals.*[17] Contact your school's Office of Student Services to learn about the resources available to you.

The purpose of this section is to describe some factors contributing to the development or the exacerbation of mental health issues in the hope of breaking the ice for further discussion between members of the academic community. Using reports from the recent literature, this

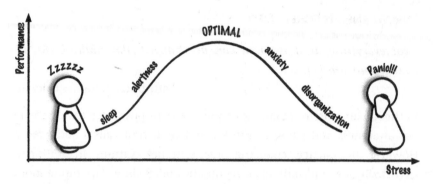

12.3 The zone: operating under a healthy amount of stress.

section highlights factors both negatively affecting distress or positively influencing mental health outlook toward a more enjoyable doctoral experience.

STRESS VS DISTRESS

Stress is a normal response to internal or environmental perturbations. Figure 12.3 illustrates how good stress, referred to as eustress, is associated with improved performance.[18] Not only will you find the motivation to run faster the day a chimpanzee chases you but your brain also marshals your heart to deliver more oxygen-rich blood to your muscles to boost your performance. Bad stress, or distress, occurs when the stress is severe, prolonged, or both.[19] Indeed, you may feel distressed if the chimpanzee decides to chase you over a very long distance (severe stress), every day (prolonged stress), or both. Distress leads to anxiety and disorganization and ultimately to diminished performance—especially if the angry chimpanzee catches you! Aside from famous primatologist Dr. Jane Goodall, few scientists deal with wild chimpanzees. Yet, "...graduate students are more than six times as likely to experience depression and anxiety as compared to the general population" (Teresa M. Evans et al., *Nature Biotechnology*[20]),

WORKPLACE CHIMPANZEES

What are modern-day chimpanzees? What are common risk factors in the workplace? Your lab, as any workplace, includes several sources of stress:[21]

work overload: graduate students often work 60+ hour weeks (this is an observation, not a recommendation!) to complete course work, lab work, and teaching duties, which represents quantitative overload. Graduate students are also at risk of qualitative overload, which occurs when complex tasks are assigned before all required skills are mastered;

lack of recognition: graduate students may seldom rely on money as a sign that their contribution is being recognized (#euphemism). Even the currency of science, aka publications and conference presentations, which eventually acknowledges one's hard work, only occurs sporadically throughout the doctoral journey, and not often enough to provide a steady stream of appreciation;

lack of control and information: when the big picture is obscure, or evaluation criteria unknown, one may feel as if they are removed from the pilot seat of their own voyage;

lack of support: completing doctoral studies taps on a broad constellation of resources from receiving proper scientific and technical advice to proper personal and social support;[22]

deadlines: graduate school is all about deadlines, lots of them stemming from parallel, yet often conflicting, tracks. Indeed, academic requirements, research milestones, and networking and outreach events must be carefully planned to avoid constantly extinguishing fires;

role ambiguity: under the somewhat loose hierarchy of academia, it is easy to lose track of who's doing what and feel responsible for decisions above one's pay grade;

work schedule: many learned from the pandemic that a flexible schedule may rapidly become a nightmare as work hours bleed into personal time; and

ethical dilemmas: understanding and following all rules of academia is already demanding, but creating them as one ventures in uncharted territories is even more strenuous.

In addition to the common workplace chimpanzees, several factors affect the mental health of graduate students. Among negative factors, a recent study further highlights:[23]

financial concerns: unemployment and job insecurity are well-known sources of mental illness,[24] to which Barreira et al. add that the

doctoral journey corresponds to a half-decade-long experience in job
insecurity,[25]

poor mentorship: advisors have a strong impact on the student's experience.[26] In addition to scientific and technological guidance, faculty
should strive to show students that they care about their success; and

discrimination: lab culture should be inclusive and welcoming; alas, some
graduate students still experience discrimination and harassment in
one form or another, turning the lab into a most unhealthy work environment. Information about campus policies (on discrimination and
harassment), reporting procedures, and counseling resources available
to students should be communicated clearly and often.

All of the situations mentioned are tricky to address for students who
may feel they embark on a conflict resembling that of David vs Goliath.
Solutions do exist: somewhere in the spectrum ranging from meeting
with a counselor from your school's Office of Student Services (for when
you doubt the gravity of the offense) to knocking on the ombudsperson's door (see definition 12.1) for situations you find unambiguously
unacceptable.

CONSEQUENCES OF STRESS

From insomnia to irritability, stress affects us in one or several facets, as
illustrated in figure 12.4. Physiological consequences include, but are not
limited to, tightness in the stomach, elevated heart rhythm and blood
pressure, heavy breathing, eating and sleeping disorders, while psychological effects include a diminished ability to deal with negative and
positive emotions. Stress and depression affect attention and other cognitive functions making it difficult to acquire and create knowledge. Stress
also influences how we deal with others, making it harder to forge and
maintain relationships.

When stress becomes distress, students may lose confidence in their
ability to finish their degree[27] and are twice as likely to leave school before
their doctoral degree is completed.[28] As a young adult, if you suffer mental health issues, you are not alone: in the US, it is the third-costliest
disease for employers.[29]

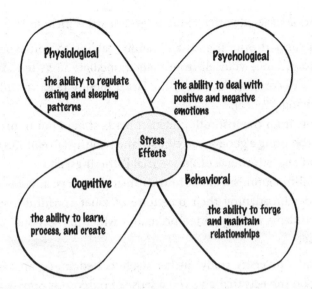

Physiological

the ability to regulate eating and sleeping patterns

Psychological

the ability to deal with positive and negative emotions

Stress Effects

Cognitive

the ability to learn, process, and create

Behavioral

the ability to forge and maintain relationships

12.4 Stress affects all spheres of our lives: physiological, psychological, cognitive, and behavioral.

STRATEGIES

Strategies to reduce workplace-related stress are articulated around three themes:

reduce: a preliminary strategy consists in reducing or eliminating stress factors. For example, providing a guaranteed stipend to graduate students may ease financial concerns;

resist: a secondary strategy consists in increasing one's resistance to stress. For example, by implementing a buddy system, departments are increasing the social support from peers. Most student services also provide workshops to learn how to cope with student life; and

treat: once mental health is compromised, seek help. Treating the soul is no more shameful than taking acetaminophen or paracetamol for a headache or crutches for a broken foot. Where to start? Most student services provide anonymous counseling and free consultation with mental health professionals. Schools openly discussing mental health strategies contribute to destigmatizing a situation affecting too large a fraction of the graduate student population.

Factors associated with a reduced level of stress include:[30]

- social support from friends and family: talking to non-experts forces you to distance yourself from your immediate tasks (e.g., code that refuses to compile) to look at the larger picture (e.g., contributing to advancing science);
- support from the institution: departments should put in place activities promoting a good social climate and healthy lab cultures; students should take advantage of these social gatherings; and
- cultivating optimism about future career prospects: advisors may also consider broadening their repertoire of what constitutes success so that students feel supported no matter which professional path they choose.

A Harvard University study further suggests hedging against failure by working on subjects that give you a sense of purpose as opposed to chasing the highest impact paper,[31] and considering participating in a side project to maximize the odds of succeeding in at least one of them. It too mentions the benefit of creating a network of peers through volunteering activities within the department, such as organizing professional development networks or social events.

Online communities are building around the theme of mental health during graduate studies. Discussion groups may be found through professional societies or common social networking platforms. Several blogs also discuss ways to promote a healthier approach to doctoral education. A common theme across such blogs is *Anima Sana In Corpore Sano* (ASICS), Latin for a sound mind in a sound body:[32]

- build downtime in your calendar: make an appointment with yourself to go for a walk, learn salsa dancing, or cook a good meal for friends;
- touch base with family and friends: yes, call a loved one!;
- practice mindfulness;
- prioritize physical as well as mental health;
- seek help when you need it: engineering students are notoriously bad at seeking help;[33] and
- remember that a PhD is a temporary situation, yet try not to let an unhealthy situation degenerate: five years is a long time to be unhappy!

12.3 CULTIVATING A SAFE PLACE

More than twenty years ago, the fictional Dr. Ross Geller—from the generation-defining sitcom *Friends*—learned that it was more than *frowned upon* for an academic to date a student and, if I might add, downright illegal when the student is underage! The power relation between mentors and mentees is very asymmetric: students depend on professors for graduating, for their careers (to get a PhD and then some great letters of reference), and to eat (through their limited stipends). Additionally, students living on campus depend on being students to keep a roof above their heads, while international students must maintain their academic status to keep their visas. Being a graduate student often implies putting very many eggs in one tiny basket.

ASYMMETRY AS AN AMPLIFIER

The lab culture is often dictated by the hierarchical structure set forth by your advisor. Some advisors operate under a vertical management style (i.e., as a distant sink-or-swim boss), while others use a more horizontal approach (i.e., as a close micro-managing coach). Student distress resulting from various vulnerabilities and management mishaps, or destructive leadership behaviors,[34] may occur in any lab culture. However, research suggests that a distant boss may amplify the anguish.[35] Indeed, imagine a (close) lab mate texting you on a Friday night: "We need to talk first thing Monday morning." You probably picture a new idea worth discussing. The exact text message from your boss, especially a distant one, triggers a different reaction. Chances are you will not sleep all that well over the weekend. Of course, managers are trained not to send such messages on Friday night. But unfortunately, professors are not formally trained in management (Can you tell?).

A close mentoring relationship with your advisor, however, does not shield you from feelings of anguish and distress. Indeed, the proximity of an extra-friendly advisor may be unsettling for students used to a strict hierarchy with well-established boundaries. The anguish may, of course, be exacerbated when such boundaries are crossed.

12.5 Relationships between faculty and students are highly asymmetrical in power. Most schools ask that they be reported such that a structure may be put in place to ensure fair and unbiased treatment for the student.

POWER IMBALANCE

Love is already complicated enough without adding power imbalance. Consensual relationships between adults of unequal power—professor and graduate student, professor and postdoc, graduate teaching assistant and undergraduate student—may lead to a conflict of interest, favoritism, and exploitation, or the perception thereof.[36] While some schools ban such relationships, many ask that they be disclosed to ensure fair treatment for the person with less power. This is not about voyeurism but about providing a framework for unbiased opinions and guidance.

Love is even more complicated when unrequited (cf. figure 12.5), especially from people in a position of authority over you and your career. This section does not offer cookie-cutter solutions to distress arising from every type of student vulnerability. There is no playbook for interpersonal relationships—now that would be a best-seller, wouldn't it? However, resources exist to help you navigate your academic life's awkward and uncomfortable circumstances. Indeed, to provide students with healthy working and learning environments, universities are constantly refining their policies to prevent violence, harassment, and incivility further.

VIOLENCE VS HARASSMENT VS INCIVILITY

Within the framework of laws against violence, most universities further deploy anti-harassment policies as part of their codes of conduct.[37] Yet, despite efforts and awareness campaigns—somewhat prompted by the #metoo movement—a recent survey still reports that 11 percent of female students experienced some form of unwanted sexual behaviors during their postsecondary studies.[38] Reporting unwanted behaviors is always daunting: to an initial injury is added the insult of having to relive the situation to explain and document it.

Academia can and must do better. And, as a member of academia, we all have a role to play in understanding, preventing, and improving our environment to make it hospitable to all. This section is not a substitute for understanding the law or your school's code of conduct and policies but a brief introduction to the topic. According to the International Labour Organization (ILO), violence and harassment in the workplace refer to a range of unacceptable behaviours and practices, or threats thereof, whether a single occurrence or repeated, that aim at, result in, or are likely to result in physical, psychological, sexual or economic harm, and includes gender-based violence and harassment. (ILO's Convention No. 190[39])

A more specific definition of harassment is "to annoy persistently; to create an unpleasant or hostile situation, especially by uninvited and unwelcome verbal or physical conduct" (Merriam-Webster[40]).

Additionally, in the workplace, incivility is defined as a "low-intensity deviant behavior with ambiguous intent to harm the target, in violation of workplace norms for mutual respect. Uncivil behaviors are characteristically rude and discourteous, displaying a lack of regard for others" (Andersson and Pearson[41]).

The following equation is common to all three situations:

$$\text{violence} = \text{target(s)} + \text{perpetrator(s)}$$

$$+ \text{ an environment condoning violence.}$$

Where can you make a difference, you ask? As an ally, you:

- ought to familiarize with your school's policies and learn about resources;

- could open dialogue with the (perceived) victim when you witness an uncomfortable situation ("are you ok?" or "do you need help?");
- may contribute to preventing a situation from happening or degenerating (e.g., make sure a colleague makes it home safely after late-night experiments or extended happy hours) or speak up when your lab environment is not as hospitable as it should be; and
- are expected to report awkward situations, incidents, or crimes you witness to someone in authority. Seek help for yourself as well: don't carry the burden of someone else's distress all by yourself.

As a victim, you:

- should attempt to remove yourself from a violent situation as soon as possible by calling the police, your school's security, or a person in authority;
- could speak to someone (a friend, someone you trust, someone from your university student office or office of violence prevention, the ombudsperson) for guidance on what to do next. Support is confidential: no one can file complaints without your explicit consent;
- may file a complaint to a person in authority (your school's office of violence prevention, your department chair, the ombudsperson, your's school security, or the police); and
- should never isolate yourself—when you are ready, seek help.

Definition 12.1: Ombudsperson   *A university ombudsperson acts as a neutral guide for community members by providing information and advice on rules and procedures, investigating complaints of unfair treatment, and recommending conflict resolution techniques, including mediation.*[42]

As the perpetrator of incivility, you must (learn to and) apologize properly by first admitting your wrongdoing, second recognizing the harm done to others, and third pledging to stop the problematic behavior.

Box 12.2
Is This Cultural?

In 2001, during orientation week at MIT, international graduate students received a briefing from a policewoman on local laws. Since laws differ

---

Box 12.2 (continued)

from country to country and even state by state, she illustrated several laws with examples. For each, she would begin with: "In the Commonwealth of Massachusetts, it is illegal to..." During the question period, one participant jokingly asked whether such terrible crimes were allowed in Rhode Island, a neighboring state. Indeed, to most international students, the acts of violence she was describing seemed universally accepted as being wrong. While the definition of harassment is also universal, it is, arguably, of less common knowledge. Most universities thus provide workshops with examples and strategies to keep their members on the right side of grace, courtesy, and the law. Incivility, which depends on subjective definitions of politeness and good manners, may be cultural. As a mild example, academic titles vary from one country to the next. In Germany, I would be *Frau Doktor Professor* Boudoux, while in Montréal, I am more often than not Caroline and sometimes Miss Boudoux, or (even worse) Madam Caro. Using inappropriate titles is on the soft side of incivility, but it might be a cause of awkwardness to someone expecting a certain title. Whatever your discomfort is, through workshops, you may learn strategies to ease your interactions with others, sometimes with humor rather than conflict. When this fails, do not hesitate to knock on the door of a person in authority or a mediator, such as the ombudsperson. Again, five years is a long time to be irritated.

PSYCHOLOGICAL SAFETY

A healthy environment also includes psychological safety, defined by Harvard Business School professor Amy C. Edmondson as the shared belief by team members that they will not be humiliated, rejected, or punished for voicing opinions and ideas.[43] As for lab culture, the advisor typically sets the tone. But as a team member, you may question whether you benefit from psychological safety and whether you contribute to creating a safe zone for others to express their creative ideas. How do you know you are in the zone? In his book *Think Again*,[44] Adam Grant provides evaluation guidelines, summarized in table 12.1.

Despite the best efforts from teammates, finding the psychologically safe zone may be a work in progress, especially if your impostor syndrome speaks louder than you do. So be patient and try to discern what factors come from the team and what factors come from within. On the other hand, if you have always found yourself in the left column of table 12.1,

Table 12.1 Benefits of psychological safety[45]

In the zone	Outside of the zone
See mistakes as a learning moment	See mistakes as a threat
Willing to gamble (and possibly fail)	Stay the course
Speak one's mind in meetings	Keep one's ideas to oneself
Openly discuss struggles	Only brag about successes
See supervisor as a (future) colleague	See supervisor as a boss

be the ally your team needs and make room for thoughts from quiet members.

12.4 THE BABY BUMP

Nothing forces you to consider work-life balance more urgently than the prospect of becoming a parent. Along with the titles professor, author, and entrepreneur, my LinkedIn profile now includes rookie mom. I did not experience motherhood as a graduate student but as a fully established faculty. I became acquainted with work-family balance and family-friendly policies in academia through colleagues during grad school and then through members of my team who became parents before I did. As a mom now, I can now discuss both sides of the medal—albeit from the point of view of an established faculty member in one of the most family-friendly provinces (Québec) and institutions (PolyMtl). Every chapter of this book includes geographic nuances, and this section on parenthood is no exception. Here, I discuss some of the conditions offered at PolyMtl not so much to brag about them but perhaps as an inspiration for future negotiations and, certainly, as an incentive for you to inquire about resources offered by your school, government, professional societies, and private foundations.

PARENTAL LEAVE
Some countries and institutions provide parental support for graduate students and postdoctoral fellows. For example, Canada's funding agencies allow paid parental leaves for up to twelve months for trainees, either receiving a scholarship or a stipend from one of its federal research

councils. Not every student, however, is paid through federal funds. Discuss your situation with your student service office or student advisor.

Twelve months is never enough, but at the same time, one may ask: *is it feasible to cut yourself entirely from research for one year?* Québec's provincial law suggests that every attempt should be made to grant you the option to do so. However, the law cannot prevent an international competitor from continuing their research and solving the question you had originally planned to study. Discuss with your advisor, find a pace that suits you, and modify that pace once you realize what parenthood is really like. If the question you had been working on before your baby was born is resolved on your return (whenever that is), a creative PI should be able to guide you toward another outstanding research question.

Remark 12.1: Other Kinds of Leaves *Life does not stop because you are enrolled in a PhD program. Your parents, loved ones, or yourself may become sick. Do not isolate yourself. Someone senior (your advisor, department head, or a counselor at your school's student office) will guide you toward the best resources and mechanisms to take time off of your studies.*

DAYCARE

Québec (and soon the rest of Canada shall follow suit) subsidizes daycare through a network of kindergartens called *centre de la petite enfance* (CPE) at a daily rate compatible with a student stipend.[46] PolyMtl further sponsors one of these CPE with priority for students' and employees' children. Many governments (though not all) offer some family-friendly measures to new parents (find out what they are); some schools also provide daycare for their community (idem). Daycare options may be offered at some conferences, and many professional societies, such as IEEE, Optica, and The International Professional Society for Optics and Photonics (SPIE), offer childcare grants to parents who bring small children to conferences (or extra help at home when children are not traveling).

Remark 12.2: Ship It *Many hotels have cribs and will allow you to have items delivered to your name before your stay. To facilitate traveling with an infant, consider using a common marketplace to ship baby supplies (diapers, formula, saline water, and so on) directly to your hotel in an attempt to travel light-ish.*

SAFETY FIRST!

Despite what we believe are family-friendly policies, many future mothers hide their pregnancies for fear of incomprehension from their PIs. Please do not. The head of the lab is responsible for the safety of everyone, born or unborn. Unbeknownst to you, your laboratory may contain several teratogenic (i.e., foetus-harming) chemicals, and the safety threshold for a pregnant woman may be as small as a few drops. As soon as you start telling your friends the happy news, discuss your pregnancy with your friendly safety officer so you can adapt your experimental work and workstation together.

Let irony win and make the *last* words of this chapter be safety *first*!

13

CONCLUDING REMARKS

"If a topic is important enough to deserve an entire book, it shouldn't end. It should be open-ended."

Adam M. Grant[1]

The ambition of this book was to share with rookie graduate students what most doctors in engineering eventually learn and perhaps the hard way. In addition to making the process more enjoyable, I hope this book opens the doors to the Ivory Tower a little wider. The questions of tomorrow will only increase in complexity and will require all *little grey cells* on deck to solve them. As a community, we cannot let ambiguities or unsaid advice get in the way of curious and talented minds contributing to the technology and science efforts.

13.1 DO AS I SAY, NOT AS I DID

I was not a perfect PhD student. My initial inertia to try things out in the lab took a while to overcome. I was terrified of submitting my first paper for publication. I paid way too much attention to acing classes. With time, I tamed the lab, understood that publishing was a process I could master, and refrained from signing up for too many lectures. Not knowing everything from the start is absolutely fine. What counts is being open to learning from every opportunity, cultivating your strengths and network, and growing out of your shortcomings.

PLAY YOUR STRENGTHS

Looking back, I can now identify some of the strengths I had as a graduate student. I learned quickly and could make friends easily. After practicing, practicing, and practicing, I could deliver a convincing oral presentation. Work-wise, I was organized and determined. Oh, and I had terrific, out-of-this-world advisors, turned mentors, turned friends. They were then rising (super)stars in the field. So much so that the mere thought of following in their footsteps was daunting. Their model was (and still is) inspiring and intimidating in equal parts. It is not just that they were incredibly smart and talented; their every effort was about pushing the limit of knowledge by everyday challenging the status quo and designing novel strategies to solve the most complex problems. I knew then that I would never be able to accomplish even a fraction of what they did. I also know now that the route they took is not the only one that leads to impact. I found mine, and I hope that this book helps you find yours.

BE THE NEXT ROLE MODEL

Sooner than later, you will become the next generation's role model. Unbeknownst to you, you probably already are a model to several undergraduate students. Welcome the next wave of engineers and scientists the way you wish the previous generation had welcomed you. Be generous with your time, and never forget how intimidating a few more years of experience may seem. Share what you know freely: don't assume that it goes without saying.

ACKNOWLEDGMENTS

This book piggybacks on material prepared for PolyMtl's doctoral workshops originally assembled by Prof. Jean Nicolas (Université de Sherbrooke, Québec, Canada), Prof. Patrick Desjardins (PolyMtl, Québec, Canada), and Dr. Yves Langhame (previous scientific director, Hydro-Québec Research Institute (IREQ), Québec, Canada). Throughout the years, I worked very closely with CAP workshop coordinators Dr. Elise St-Jacques and Dr. Minea Valle-Frajer to curate the list of topics presented here. Funding for researchers for this book, including Dr. Felipe Gohring de Magalhães and Ms. Wanessa Cardoso de Sousa, was partly provided by PolyMtl's Office of pedagogical support and innovation, from the French *Bureau d'appui et d'innnovation pédagogiques* (BAIP) and the Office of Educational Support and Student Experience. I also want to acknowledge the many doctoral candidates whom I have had the pleasure to teach for their generous feedback.

I typically write books with tons of equations, leaving little room for style or interpretation. This guide is different. While primarily based on literature, it is tainted by my heuristics as a PhD candidate and, now, as a mentor. Despite two decades in the field, several dead angles remained. I am grateful for several colleagues and friends (victims?) listed as contributors (see p. 305) who have generously reviewed and enhanced this manuscript. In particular, I am immensely indebted to Prof. Jason R. Tavares for his thorough critique of each chapter and his enthusiastic, thoughtful, and humorous additions to the work. Let this be the beginning of a long and fruitful collaboration!

Speaking of wonderful collaborations, many thanks are due to Dr. Jermey Matthews, Haley Biermann, Jitendra Kumar, and the fantastic team of copy editors at The MIT Press Press who have finally taught a franco-Belgo-Canadian the proper use of articles in (the) English. As a teacher (and researcher), I was inspired early on by Professor Eric Mazur's work and was very touched by his foreword. And as a die-hard fan, I was delighted that Tom Gauld agreed to illustrate the cover.

I finally want to thank my parents, Francine and Michel, for being models of curiosity, passion and hard work. They used to say that "impossible n'est pas français," which translates into the word impossible not belonging to the French vocab. Later, during my PhD, under the guidance of Profs. Brett E. Bouma and Guillermo J. Tearney, I came to realize that the word impossible was equally unwelcome in their lab. Eventually, I made my own contributions to knowledge and guided graduate students toward theirs. And just when I thought I had some things figured out, I came across opportunities that pulled me outside of my comfort zone. I am grateful for my son, Émile, who teaches me the true meaning of curiosity, for my close friends, who help smoothen many learning curves, and for my partner, Mario, who, every day, contributes to domesticating my impostor syndrome. *Merci du fond du cœur.*

ABBREVIATIONS

A/V audio-visual. 221
AAAS American Association for the Advancement of Science. 271, 276
AI artificial intelligence. 28, 128, 152, 203
APL Applied Physics Laboratory. 112
ASICS *Anima Sana In Corpore Sano.* 256
AUC area under the curve. 24

BAIP Office of pedagogical support and innovation, from the French *Bureau d'appui et d'innnovation pédagogiques.* 267
BIPCV Office of intervention and prevention of conflicts and violence, from the French *Bureau d'intervention et de prévention des conflits et de la violence.* 283
BIPOC Black, Indigenous, and people of color. 152, 246
BK7 borosilicate glass. 237

CAP researcher, actor of progress, from the French *chercheur, acteur de progrès.* 267
CBS Columbia Broadcasting System. 149
CIHR Canadian Institutes of Health Research. 277
CMS citation management software. 80, 81

CNC computer numerical control. 168
CPE *centre de la petite enfance.* 263
CSA Canadian Space Agency. 97
CT computed tomography. 240
CV curriculum vitae. 105, 108, 109, 114, 116, 117, 276

DAST draw-a-scientist test. 246–248
DOD US Department of Defense. 156, 158
DOE US Department of Energy. 146
DORA Declaration on Research Assessment. 30

f2f finish-to-finish. 164
f2s finish-to-start. 164
FIFA *Fédération Internationale de Football* Association. 3

GFP green fluorescent protein. 238
GPA grade point average. 50
GPS global positioning system. 111, 112

HHMI Howard Hughes Medical Institute. 130, 186
HQP highly qualified personnel. 118
HR human resource. 113, 276

IAP independent activity period. 50, 272
ICC International Cricket Coucil. 3
IDEA inclusion, diversity, equity, and accessibility. 152, 247, 281, 284
IEEE The Institute of Electrical and Electronics Engineers. 65, 263, 273
IF impact factor. 29, 30, 271
ILO International Labour Organization. 259
IP intellectual property. 28, 45, 97, 100, 127, 132, 137, 182, 183, 190, 197, 201, 204, 207, 223, 225, 226, 228, 230
IPO initial public offering. 276
IRB institutional review board. 201, 202
ISS International Space Station. 159

s2f   start-to-finish. 164
s2s   start-to-start. 164
SECM   spectrally encoded confocal microscopy. 90
SNR   signal-to-noise ratio. 89, 109, 242
SO   specific objective. 70, 73, 86, 87, 90, 91
SOs   specific objective. 86, 156–162, 167
SPIE   The International Professional Society for Optics and Photonics. 263
SSHRC   Canada's Social Sciences and Humanities Research Council. 277
STEM   science, technology, engineering, and mathematics. xxii, 1, 6, 9, 98, 102, 115, 246, 249

TA   teaching assistantship. 34, 62
TED   Technology, Entertainment, Design. 59, 107
TRL   technology readiness level. 125, 126, 225
TTO   technology transfer office. 111, 228–230
TV   television. 97, 275

UK   United Kingdom. 218, 220, 270, 272–274
UN   United Nations. 151
US   United States of America. 98, 146, 169, 218, 219, 254, 270, 272, 276, 277, 282
USPTO   United States Patent and Trademark Office. 227

WBS   work-breakdown structure. 94, 156, 158–162, 165
WP   work package. xiv, 156, 158–163

NOTES

PREFACE

1. Hergé, *Les bijoux de la Castafiore* (Belgium: Casterman, 1963), 62.

2. Aka CAP workshops.

CHAPTER 1

1. Richard Phillips Feynman, *"Surely You're Joking, Mr. Feynman!": Adventures of a Curious Character* (W. W. Norton, 1985).

2. Catherine Crouch et al., "Classroom Demonstrations: Learning Tools or Entertainment?," *American Journal of Physics* 72, no. 6 (2004): 835–838.

3. Catherine G. P. Berdanier et al., "Analysis of Social Media Forums to Elicit Narratives of Graduate Engineering Student Attrition," *Journal of Engineering Education* 109, no. 1 (2020): 125–147.

4. Sergio Leone, *The Good, the Bad and the Ugly*, Produzioni Europee Associate, 1966.

5. Feynman, *"Surely You're Joking, Mr. Feynman!"*

6. Source: payscale.com.

7. Lecture series on Scott Galloway, *The Algebra of Happiness: Notes on the Pursuit of Success, Love, and Meaning* (New York: Penguin, 2019).

8. Douglas Adams, The Hitchhiker's Guide to the Galaxy, Pan Books, 1979.

9. Margaret F. King, *Ph.D. Completion and Attrition: Analysis of Baseline Demographic Data from the Ph.D. Completion Project* (Conroe, TX: Nicholson, 2008).

10. Vincent Larivière, "PhD Students' Excellence Scholarships and Their Relationship with Research Productivity, Scientific Impact, and Degree Completion," *Canadian Journal of Higher Education* 43, no. 2 (2013): 27–41.

11. John Batelle, "The Birth of Google," Wired, August 1, 2005, https://www.wired.com/2005/08/battelle/.

12. Katie Langin, "It's OK to Quit Your Ph.D.," *Science*, June 25, 2019. https://doi.org .10.1126/science.caredit.aay5196.

13. The Wachowskis, *The Matrix*, Warner Bros., 1999.

14. Peter A. Daempfle, "An analysis of the High Attrition Rates Among First Year College Science, Math, and Engineering Majors," *Journal of College Student Retention: Research, Theory & Practice* 5, no. 1 (2003): 37–52.

15. Brandi N. Geisinger and D. Raj Raman, "Why They Leave: Understanding Student Attrition from Engineering Majors," *International Journal of Engineering Education* 29, no. 4 (2013): 914–925.

16. Sarah Jane Bork and Joi-Lynn Mondisa, "Engineering Graduate Students' Mental Health: A Scoping Literature Review," *Journal of Engineering Education* 111, no. 3 (2022): 665–702.

17. Juan M. Cruz et al., "Revising the Dissertation Institute: Contextual Factors Relevant to Transferability," (ASEE Annual Conference & Exposition, 2019).

18. Ellen Zerbe et al., "Engineering Graduate Students' Critical Events as Catalysts of Attrition," *Journal of Engineering Education* 111, no. 4 (2022): 868–888; Berdanier et al., "Analysis of Social Media Forums to Elicit Narratives of Graduate Engineering Student Attrition"; Solveig Cornér, Erika Löfström, and Kirsi Pyhältö, "The Relationship Between Doctoral Students' Perceptions of Supervision and Burnout," *International Journal of Doctoral Studies* 12 (2017): 91–106, https://doi.org/10.28945/3754.

19. National Science Foundation, "Survey of Earned Doctorates," 2021, https://ncses .nsf.gov/pubs/nsf22300/data-tables.

20. Michael T. Nietzel, "Ten Ways U.S. Doctorates Have Changed in the Past 20 Years," Forbes.com, 2021, https://www.forbes.com/sites/michaeltnietzel/2021/10 /13/ten-ways-us-doctoral-degrees-have-changed-in-the-past-20-years/?sh=1dd71416 2a71.

21. Patrick Desjardins, "CAP7003E Doctoral Strategies in Engineering," class notes.

22. Class notes from Jean Nicolas, *Réussir mon doctorat*, 2012; class notes from Yves Langhame, "CAP7015E Leading a Research Project." 2015.

23. The original joke named two famous schools in the Boston metropolitan area, but let me paraphrase it so as to not reinforce stereotypes. "MIT Can't Read or Harvard Can't Count Joke," The Los Angeles Times, October 23, 1966 as reported by as reported by https://www.newspapers.com/clip/77814429/mit-cant-read-or-harvard -cant-count/.

24. My *artwork* requires a legend: who wears glasses is an advisor; who does not wear glasses is a student. Gray hair was harder to draw in black and white.

CHAPTER 2

1. Despite what the name implies, a postdoctorate, which comes after the PhD, is not a university degree but the title of an intermediate position between the doctorate and a first faculty appointment.

2. A doctoral mentor is interchangeably called an advisor in the US or a supervisor in the UK.

3. I am, however, a Dr. abroad, which may explain why I travel so much.

4. Matt Might, *The Illustrated Guide to a Ph.D.* (November 2020), http://matt.might .net/articles/phd-school-in-pictures/.

5. Association du Corps Intermédiaire de l'EPFL, *Practical Guide for PhD Candidates at EPFL*, (December 2011), http://acide.epfl.ch/wp-content/uploads/2014/12/PhD _Guide_2011.pdf.

6. The Guardian, "Doorbell Cam Captures Moment Paul Milgrom Finds Out He Has Won the Nobel Prize for Economics," YouTube video, 02:01, https://www.youtube .com/watch?v=JhfDyBLRnrM.

7. From Destin Daniel Cretton, *Shang-Chi and the Legend of the Ten Rings*, (Marvel Studios, 2021).

The author—nowhere cool enough to watch Marvel's marvels—would like to acknowledge the kind contribution of Prof. Jason R. Tavares for his generous supply of inspirational quotes and humorous bits.

8. Charles X. Ling and Qiang Yang, "Crafting Your Research Future: A Guide to Successful Master's and Ph.D. Degrees in Science & Engineering," *Synthesis Lectures on Engineering* 7, no. 3 (2012): 1–168.

9. Barbara E. Lovitts, *Making the Implicit Explicit: Creating Performance Expectations for the Dissertation* (Stylus Publishing, 2007).

10. Lovitts, Making the Implicit Explicit.

11. From Prof. Patrick Desjardins's slides. PolyMtl's CAP7003E *Doctoral Research Strategies in Engineering*.

12. Paul Adrien Maurice Dirac, Quantum Mechanics, PhD Thesis (Cambridge: Cambridge University press, 1926).

13. Maria Goeppert Mayer won the 1963 Nobel prize in physics for work related to her doctoral dissertation published in: Maria Goeppert Mayer, "Elementary Processes with Two Quantum Jumps," *Annalen der Physik* 9 (1931): 273–294.

14. W. Kaiser and C. G. B. Garrett, "Two-Photon Excitation in Ca F 2: Eu 2+," *Physical Review Letters* 7, no. 6 (1961): 229.

15. Winfried James H. Strickler Denk and Watt W. Webb, "Two-Photon Laser Scanning Fluorescence Microscopy," Science 248, no. 4951 (1990): 73—76.

16. In most aspects, I wish for this book to age well. Here, however, I wish for the rapid obsolescence of such a small figure.

17. Alan M. Turing, "Computing Machinery and Intelligence," *Mind*, no. 49 (1950): 433–460.

18. Heather Cray, "How to Make an Original Contribution to Knowledge," *University Affairs/Affaires Universitaires*, August 2014.

19. Philip Abraham et al., "Duplicate and Salami Publications," *Journal of Postgraduate Medicine* 46, no. 2 (2000): 67.

20. Joan M. Reitz, *Online Dictionary for Library and Information Science* (Danbury, CT: Western Connecticut State University, 1996).

21. The IF is calculated by the Journal Citation Reports, operated by Clarivate Analytics.

22. Vincent Lariviere and Cassidy R. Sugimoto, "The Journal Impact Factor: A Brief History, Critique, and Discussion of Adverse Effects," in *Springer Handbook of Science and Technology Indicators*, ed. Wolfgang Glänzel et al. (Springer, 2019), 3–24.

Editorial, "Time to Remodel the Journal Impact Factor," *Nature* 535, no. 466 (2016).

23. Vincent Lariviere et al., "A Simple Proposal for the Publication of Journal Citation Distributions," *BioRxiv*, 2016, 062109.

24. American Society for Cell Biology et al., "San Francisco declaration on research Assessment (DORA)," 2012.

25. Peter Suber, *Open Access* (MIT Press, 2012).

26. Bo-Christer Sari Kanto-Karvonen Björk and J. Tuomas Harviainen, "How Frequently Are Articles in Predatory Open Access Journals Cited," *Publications* 8, no. 2 (2020): 17.

27. Renowned publishing houses include Springer Nature (publishing *Nature*), Elsevier, Wiley, MIT Press, and so on.

28. Technical societies include the American Association for the Advancement of Science (AAAS) (publishing Science), IEEE, Optica, and so on.

29. Remember, however, that these lists are never complete as new titles sprout on a regular basis.

30. *Think.Check.Submit.* https://thinkchecksubmit.org/journals, December 2020.

31. Jorge E. Hirsch, "An Index to Quantify an Individual's Scientific Research Output," *Proceedings of the National Academy of Sciences* 102, no. 46 (2005): 16569–16572.

32. G. Conroy, "What's Wrong with the H-index, According to Its Inventor," in *Nature Index*, March 24, 2020.

33. Richard Van Noorden and Dalmeet Singh Chawla, "Hundreds of Extreme Self-Citing Scientists Revealed in New Database," *Nature* 572, no. 7771 (2019): 578–580.

34. Diana Hicks et al., "Bibliometrics: The Leiden Manifesto for Research Metrics," *Nature* 520, no. 7548 (2015): 429–431.

35. Jorge E. Hirsch, "Does the H Index Have Predictive Power?," *Proceedings of the National Academy of Sciences* 104, no. 49 (2007): 19193–19198.

36. Not that Newton needs it to promote his career anymore.

37. She was indeed initially denied a faculty position, but for reasons that had nothing to do with her h-index: at the time, husband and wife could not serve the same department. Eugene P. Wigner, "Maria Goeppert Mayer," *Physics Today* 25, no. 5 (May 1972): 77–79.

38. Evan Tarver, Corporate Culture, 2020. https://www.investopedia.com/terms/c/corporate-culture.asp.

39. Merriam-Webster staff, Merriam-Webster's collegiate dictionary, vol. 2 (Merriam-Webster, 2004).

40. Els van Rooij, Marjon Fokkens-Bruinsma, and E. Jansen, "Factors that Influence PhD Candidates' Success: The Importance of PhD Project Characteristics," *Studies in Continuing Education* 43, no. 12 (2019): 1–20.

41. Roel Snieder and Ken Larner, *The Art of Being a Scientist: A Guide for Graduate Students And Their Mentors* (Cambridge: Cambridge University Press, 2009).

42. Jorge Cham, *The Thesis Committee*, 2012.

43. Snieder and Larner, *The Art of Being a Scientist: A Guide for Graduate Students And Their Mentors*.

44. Lovitts, *Making the Implicit Explicit: Creating Performance Expectations for the Dissertation*.

45. Breach of ethics and university codes are examples of conducts that can lead to termination.

CHAPTER 3

1. In some countries, such as Australia, a PhD is a higher degree by research only: no course work, no doctoral exam, and no thesis defense. The dissertation is solely evaluated by a committee composed of Australian and international experts https://ielanguages.com/blog/phd-in-australia/.

2. In some schools, such as Stanford University in the US, the doctoral thesis is defended before the dissertation is written. Inquire early regarding your school's doctoral journey.

3. Many variations exist to assess a candidate's preparedness for doctoral studies.

4. In some institutions, such as Danmarks Tekniske Universitet, corrections are made before the thesis defense.

5. As long as you fill out the appropriate form!

6. MIT has a lesser-known fourth session during the month of January called independent activity period (IAP) and used for flexible teaching and learning. Amongst classes offered was one called Charm School in which students learn *les bonnes manières*, or proper manners. https://news.mit.edu/2015/how-become-charmer-mit-charm-school-0213.

7. Jon Bentley, Programming Pearls (Addison-Wesley Professional, 2016).

8. This also applies to endlessly polishing the fonts and figures for a presentation. Save time by using a template.

9. Gordon Research Conferences topics can be found at: https://www.grc.org.

10. In some countries, such as in the UK, students are not required to prepare a doctoral exam unless they upgrade from a master's to a PhD.

11. In some countries, such as in the UK, students join an already funded research project with predefined goals. The concept of a thesis proposal is reserved for humanities. However defined a project is before recruitment, a PhD student should nonetheless be able to articulate its objectives, impact, and novelty, even if such a document is not a strict degree requirement. Despite not having to complete a doctoral exam, Australian students must prepare a thesis proposal and present it to a supervisory panel composed of, at least, a primary supervisor and an associate supervisor. https://policies.anu.edu.au/ppl/document/ANUP_012813 and https://www.uts.edu.au/research-and-teaching/graduate-research/supervisors-and-faculty/supervisor-panel.

12. From https://www.grad.ubc.ca/faculty-staff/policies-procedures/comprehensive-examination.

13. Gregory S. Patience, Daria C. Boffito, and Paul Patience, *Communicate Science Papers, Presentations, and Posters Effectively* (Academic Press, 2015).

14. Stephen King, *On Writing: A Memoir of the Craft* (Simon & Schuster, 2000).

15. Lariviére, "PhD Students' Excellence Scholarships and Their Relationship with Research Productivity, Scientific Impact, and Degree Completion."

16. Abraham H. Maslow, "A Theory of Human Motivation," Psychological Review 50, no. 4 (1958): 370–396. https://doi.org/10.1037/h0054346.

17. Umberto Eco, *How to Write a Thesis* (MIT Press, 2015).

18. Aimed primarily at doctoral students in the humanities, Eco's book describes sources, rather than resources.

19. Snieder and Larner, *The art of being a scientist: A guide for graduate students and their mentors*, p. 55.

20. Snieder and Larner, *The Art of Being a Scientist: A Guide for Graduate Students and Their Mentors.*

21. Jared Diamond, *Guns, Germs, and Steel* (W. W. Norton, 1997).

22. Yuval Noah Harari, *Sapiens* (Bazarforlag AS, 2016).

23. Robert A. Day and Barbara Gastel, *How to Write and Publish a Scientific Paper*, 8th ed. (Cambridge: Cambridge University Press, 2016).

24. Burroughs Wellcome Fund and Howard Hughes Medical Institute, *Making the Right Moves. A Practical Guide to Scientific Management for Postdocs and New Faculty* 2006.

25. Kathy Barker, *At the Helm: A Laboratory Navigator* (Cold Spring Harbor, NY: Cold Spring Harbor Laboratory Press, 2002).

26. Caroline Boudoux, "Wavelength Swept Spectrally Encoded Confocal Microscopy for Biological and Clinical Applications" (PhD diss., MIT, 2007).

27. Seok-Hyun Yun et al., "High-Speed Wavelength-Swept Semiconductor Laser with a Polygon-Scanner-Based Wavelength Filter," *Optics Letters* 28, no. 20 (2003): 1981–1983.

28. Caroline Boudoux et al., "Rapid Wavelength-Swept Spectrally Encoded Confocal Microscopy," *Optics Express* 13, no. 20 (2005): 8214–8221.

29. Caroline Boudoux et al., "Spectral Encoding: A Novel Platform for Endoscopy and Microscopy," in *Laser Science* (Optica Publishing Group, 2006), JWC5, https://opg.optica.org/conference.cfm?meetingid=69&yr=2006#JWC.

30. Caroline Boudoux et al., "Optical Microscopy of the Pediatric Vocal Fold," *Archives of Otolaryngology–Head & Neck Surgery* 135, no. 1 (2009): 53–64.

31. Seok-Hyun Yun et al., "Extended-Cavity Semiconductor Wavelength-Swept Laser for Biomedical Imaging," *IEEE Photonics Technology Letters* 16, no. 1 (2004): 293–295.

32. Ronit Yelin et al., "Multimodality Optical Imaging of Embryonic Heart Microstructure," *Journal of Biomedical Optics* 12, no. 6 (2007): 064021–064021.

33. Dongkyun Kang et al., "Combined Reflection Confocal Microscopy and Optical Coherence Tomography Imaging of Esophageal Biopsy," *Gastrointestinal Endoscopy* 69, no. 5 (2009): AB368.

34. IEEE defines conditions for authorship in https://journals.ieeeauthorcenter.ieee.org/become-an-ieee-journal-author/publishing-ethics/ethical-requirements/.

35. But variations exist whereby lab directors either sign first or second. Never assume: discuss the order with your advisor.

36. This is an instance where I cannot avoid the *senior* epithet.

37. Lovitts, *Making the Implicit Explicit*.

38. In some countries, such as in the UK, the thesis defense, or *viva voce*, is held *in camera* (i.e., behind closed doors). Attending one is not an option, but organizing a mock defense with colleagues who have witnessed one certainly is.

In some other countries, such as in Australia, the *viva voce* is not required (see https://insiderguides.com.au/doing-your-phd-in-australia-guide/), although some institutions are beginning to incorporate such a milestone in their doctoral journey.

CHAPTER 4

1. Eco, *How to Write a Thesis*.

2. In some countries, like in the UK, engineering students are exempt from submitting a thesis proposal. This chapter may still be of interest. Indeed, a thesis proposal and a doctoral dissertation share many traits: from the literature search to the demonstration of novelty and impact, students may use this chapter to learn how to articulate the core characteristics of their research project. Tips included here apply beyond the proposal to the thesis itself.

3. Inspired from Columbia University tutorial "How to Write a Thesis Proposal," accessed October 16, 2023, https://www.ldeo.columbia.edu/%7Emartins/sen_res /how_to_thesis_proposal.html.

4. Eco, *How to Write a Thesis*.

5. Known as small cakes best eaten outside the lab to avoid crumbs ruining your experiment.

6. Inspired from Anna Clemens, "How to Write a Title for Your Research Paper," *Dr. Anna Clemens* (blog), accessed October 16, 2023, https://www.annaclemens.com /blog/ten-most-common-mistakes-when-choosing-a-paper-title.

7. Carlos Eduardo Paiva, João Paulo da Silveira Nogueira Lima, and Bianca Sakamoto Ribeiro Paiva, "Articles with Short Titles Describing the Results Are Cited More Often," *Clinics* 67, no. 5 (2012): 509–513.

8. David V. Thiel, Research Methods for Engineers (Cambridge: Cambridge University Press, 2014).

9. Chris Hart, *Doing a Literature Review: Releasing the Research Imagination*, 2nd ed. (Thousand Oaks, CA: Sage Publications, 2018).

10. Andrew Booth, Anthea Sutton, and Diana Papaioannou, *Systematic Approaches to a Successful Literature Review*, 2nd ed. (London: Sage, 2016).

11. Zohre Momenimovahed et al., "Ovarian Cancer in the World: Epidemiology and Risk Factors," *International Journal of Women's Health* 11 (2019): 287.

12. Robert C. Bast, "Status of Tumor Markers in Ovarian Cancer Screening," *Journal of Clinical Oncology: Official Journal of the American Society of Clinical Oncology* 21, no. 10 (May 2003): 200s–205s.

13. Britt K. Erickson, Michael G. Conner, and Charles N. Landen Jr. "The Role of the Fallopian Tube in the Origin of Ovarian Cancer," *American Journal of Obstetrics and Gynecology* 209, no. 5 (2013): 409–414

14. Charles Mayo Goss, "Gray's Anatomy of the Human Body," *Academic Medicine* 35, no. 1 (1960): 90.

15. G. J. Tortora and B. Derrickson, *Tortora's Principles of Anatomy and Physiology*, 15th ed. Wiley, 2017.

16. For everything else, feel free to be like him: he was exceptional!

17. Keep in mind, however, that if you are not paying for a web-based service, you might be the product!

18. Some websites may also send you notifications of newly published work from authors you follow.

19. Luc Van Campenhoudt, Jacques Marquet, and Raymond Quivy, *Manuel de recherche en sciences sociales*, 5th ed. (Dunod, 2017).

20. If you need to look up this citation, may I respectfully suggest that you allow yourself a short break to peruse some classics from the non scientific literature? John Ronald Reuel Tolkien, *The Lord of the Rings* (Allen & Unwin, 1954).

21. George T. Doran et al. "There's a SMART Way to Write Management's Goals and Objectives," *Management Review* 70, no. 11 (1981): 35–36; Graham Yemm, *Financial Times – Essential Guides to Leading Your Team: How To Set Goals, Measure Performance And Reward Talent* (Pearson UK, 2012).

22. Matin Durrani, "Ig Nobels Prove to be More than a Joke," *Physics World* 16, no. 4 (2003): 12.

23. Boudoux et al., "Rapid Wavelength-Swept Spectrally Encoded Confocal Microscopy."

24. Mathias Strupler et al., "Rapid Spectrally Encoded Fluorescence Imaging Using a Wavelength-Swept Source," *Optics Letters* 35, no. 11 (2010): 1737–1739.

25. David C. Van Aken andWilliam F. Hosford, Reporting Results: A Practical Guide for Engineers and Scientists (Cambridge: Cambridge University Press, 2008).

CHAPTER 5

1. *Chicago Tribune* columnist Mary Schmich wrote a hypothetical commencement speech titled "Wear Sunscreen." It was later turned into a spoken-word song by Baz Luhrmann.

   Mary Schmich, "Advice, Like Youth, Probably Just Wasted on the Young (Wear Sunscreen)," *Chicago Tribune, 1997.*

2. Recipient of the 2017 Stephen Hawking Medal for Science Communication, Dr. Neil deGrasse Tyson wrote several books and columns on astrophysics.

3. Prof. Brian Cox is a professor of particle physics at the University of Manchester, in addition to being the presenter of science TV shows and having authored several popular science books.

4. Prof. Brian Greene is a professor of physics and mathematics at Columbia University and the author of books on the fabric of the universe.

5. Dr. David Suzuki is an emeritus professor at the University of British Columbia and has for more than four decades hosted radio and TV programs.

6. Even though in Québec, PhDs do not use the title Dr., here, it is appropriate as Dr. Saint-Jacques also holds an MD.

7. Jonathan Horsley, "Tom Morello Was in College Rock Band with 2022 Nobel Prize Winner in Chemistry – and They Were Good." MusicRadar, October 7, 2022, https://www.musicradar.com/news/tom-morello-rage-against-the-machine-carlolyn-r-bert ozzi-nobel-prize.

8. J. Opsomer et al., US Employment Higher in the Private Sector than in the Education Sector for US-Trained Doctoral Scientists and Engineers: Findings from the 2019 Survey of Doctorate Recipients. (NSF 21-319, 2021).

9. Like Queen's Dr. May, Bad Religion's Dr. Greg Gaffin also earned a doctorate (in zoology).

10. Maximiliaan Schillebeeckx, Brett Maricque, and Cory Lewis, "The Missing Piece to Changing the University Culture," *Nature Biotechnology* 31, no. 10 (2013): 938–941.

11. Peter Fiske, "For Your Information," *Nature* 538, no. 7625 (2016): 417–418.

12. Catherine H. Crouch and Eric Mazur, "Peer Instruction: Ten Years of Experience and Results," *American Journal of Physics* 69, no. 9 (2001): 970–977; Carla C. Johnson et al., *Handbook of Research on STEM Education* (London: Routledge, 2020).

13. Though money, trophies, and the occasional pat on the back are appreciated.

14. David M. Giltner, *It's a Game Not a Formula: How to Succeed as a Scientist Working in the Private Sector* (SPIE Press, 2021).

15. Peter S. Fiske, *Put Your Science to Work: The Take-Charge Career Guide for Scientists* (John Wiley & Sons, 2013).

16. Class notes from Jean Nicolas, *Réussir mon doctorat*, (2012).

16. Frances Trix and Carolyn Psenka, "Exploring the Color of Glass: Letters of Recommendation for Female and Male Medical Faculty," *Discourse & Society* 14, no. 2 (2003): 191–220.

17. David M. Giltner, *Turning Science into Things People Need: Voices of Scientists Working in Industry* (Wide Media Group, 2017).

18. Prof. Haushofer CV of failures is published at: https://www.sciencealert.com/why -creating-a-cv-of-failures-is-good-Princeton-professor-viral.

It was based on a piece published in *Nature* by Dr. Melanie I. Stefan Melanie Stefan, "A CV of Failures," *Nature* 468, no. 7322 (2010): 467.

19. Brownie points when an email addressed to me does not start with *Dear Sir*!

20. Not necessarily in that order.

21. Do not worry if the conversation does not take off, maybe it is they who are out of practice!

22. In the business world, a unicorn is an elusive company that reaches a valuation of one billion dollars before its initial public offering (IPO).

23. William H. Guier and George C. Weiffenbach, "Genesis of Satellite Navigation," *Johns Hopkins APL Technical Digest* 18, no. 2 (1997): 179.

24. Dr. Emmett L. Brown is a fictional scientist from the popular movie franchise *Back to the Future*. His first name is Emit, or time spelled backward. Sir Isaac Newton should not require an introduction.

25. Unless that question is: Who is Isaac Newton?

26. Devora Zack, *Networking for People Who Hate Networking: A Field Guide for Introverts, the Overwhelmed, and the Underconnected* (Berrett-Koehler Publishers, 2019).

27. The right-hand side for name tags and the *right* side up for lanyards.

28. From family members, that is. Not from HR; that would be troublesome indeed.

29. Kendall Powell, "How to Sail Smoothly from Academia to Industry," *Nature* 555, no. 7697 (2018).

30. Popular websites advertising faculty positions include: AAAs's ScienceCareers at https://jobs.sciencecareers.org/ and Nature Careers at https://www.nature.com /naturecareers.

31. https://engineeroxy.com lists academic vacancies worldwide in engineering. Other such lists exist in science or other specific fields. Subscription is often free.

32. Shulamit Kahn and Donna K. Ginther, "The Impact of Postdoctoral Training on Early Careers in Biomedicine," *Nature Biotechnology* 35, no. 1 (2017): 90–94.

33. Yes, soon you may be entitled to some paid vacation days!

CHAPTER 6

1. John A. Sharp, John Peters, and Keith Howard, *The Management of a Student Research Project* (Gower Publishing, 2012).

2. The Project Management Institute trains the next generation of project managers. https://www.pmi.org/.

3. Some European countries have shorter (e.g., two to three years). Some countries, such as Canada and the US, have longer PhDs (e.g., 3.5-5.5 years).

4. David I. Cleland and William R. King, *Project Management Handbook* (New York: Van, 1988).

5. Crane, David, and Martha Kauffman, *Friends*, Warner Bros., aired 1994–2004 on NBC.

6. Mihály Héder, "From NASA to EU: The Evolution of the TRL Scale in Public Sector Innovation," *The Innovation Journal* 22, no. 2 (2017): 1–23.

7. Project Management Institute.

8. Joan Guberman, Judith Saks, Barbara Shapiro, and Marion Torchia, Fund and Institute, *Making the Right Moves. A Practical Guide to Scientific Management for Postdocs and New Faculty*, 2nd ed., The Howard Hughes Foundation and Boroughs Welcome Fund, 2006.

9. Roger Atkinson, Project Management: Cost, Time and Quality, Two Best Guesses and a Phenomenon, Its Time to Accept Other Success Criteria, *International Journal of Project Management* 17, no. 6 (1999): 337–342.

10. Langhame, "CAP7015E Leading a Research Project."

CHAPTER 7

1. Tim Donnelly, "Brilliant Inventions Made by Mistake," *Inc.* 24 (2012), https://www.inc.com/tim-donnelly/brilliant-failures/9-inventions-made-by-mistake.html.

2. Thiel, *Research Methods for Engineers*.

3. Van Campenhoudt, Marquet, and Quivy, *Manuel de recherche en Sciences Sociales*, 5th ed.

4. Benjamin Brewster 1882 February, The Yale Literary Magazine, Conducted by the Students of Yale College, Volume 47, Number 5, Portfolio: Theory and Practice by Benjamin Brewster, Quote Page 202, New Haven, Connecticut.

5. Commercial quotes are typically attached to the grant proposal to evaluate the project's feasibility but are long expired by the time the money is granted.

6. Lovitts, *Making the Implicit Explicit*.

7. Inspired by one of Scott Adam's most delicious Dilbert strips published on August 25, 1995.

8. In Canada, for example, funding comes from three research councils: Canada's Social Sciences and Humanities Research Council (SSHRC), Canadian Institutes of Health Research (CIHR), and Canada's Natural Science and Engineering Research Council (NSERC). Together, these three funding agencies span all five disciplines.

9. Bernard C. K. Choi and Anita W. P. Pak, "Multidisciplinarity, Interdisciplinarity, and Transdisciplinarity in Health Research, Services, Education and Policy: 2. Promotors, Barriers, and Strategies of Enhancement," *Clinical and Investigative Medicine* 30, no. 6 (2007): E224–E232.

10. Xucheng Hou et al., "Lipid Nanoparticles for mRNA Delivery," *Nature Reviews Materials* 6, no. 12 (2021): 1078–1094.

11. Hou et al., "Lipid Nanoparticles for mRNA Delivery."

12. Alan Colin Brent, "Transdisciplinary Approaches to Engineering R&D: Importance of Understanding Values and Culture," in *Handbook of Sustainable Engineering*, ed. Joanne Kauffman and Lee Kun-Mo (Springer, 2013).

13. David Hockney and Charles M. Falco, "Optical Insights into Renaissance Art," *Optics and Photonics News*, no. 7 (July 2000): 52–59.

14. Choi and Pak, "Multidisciplinarity, Interdisciplinarity, and Transdisciplinarity in Health Research."

15. Funding agencies abroad have similar classifications. For example, life-science research in the US is funded through the US National Institutes of Health (NIH), while natural sciences and engineering research is funded through the NSF. Each agency manages several programs specializing in individual research fields.

16. Sharp, Peters, and Howard, *The Management of a Student Research Project.*

17. Cecilia Jarlskog, "Lord Rutherford of Nelson, His 1908 Nobel Prize in Chemistry, and Why He didn't Get a Second Prize," *Journal of Physics: Conference Series* 136, no. 1 (2008): 012001.

18. J. P. giveni Von Der Weid et al., "On the Characterization of Optical Fiber Network Components with Optical Frequency Domain Reflectometry," *Journal of Lightwave Technology* 15, no. 7 (1997): 1131–1141.

19. Yun et al., "High-Speed Wavelength-Swept Semiconductor Laser."

20. Martin Poinsinet de Sivry-Houle et al., "All-Fiber Few-Mode Optical Coherence Tomography Using a Modally-Specific Photonic Lantern," *Biomedical Optics Express* 12, no. 9 (2021): 5704–5719.

21. Breanna Bishop, "Lawrence Livermore National Laboratory Achieves Fusion Ignition," Lawrence Livermore National Laboratory, December 14, 2023, https://www.llnl.gov/news/national-ignition-facility-achieves-fusion-ignition.

22. A. S. Eddington, "The Internal Constitution of the Stars," *Nature* 106, no. 2653 (1920): 14–20, https://doi.org/10.1038/106014a0, https://doi.org/10.1038/106014a0.

23. John M. Dudley, "Defending Basic Research," *Nature Photonics* 7, no. 5 (2013): 338–339.

24. Vanevar Bush, *Science, the Endless Frontier* (Washington: National Science Foundation–EUA, 1945).

25. Donald E. Stokes, *Pasteur's Quadrant: Basic Science and Technological Innovation* (Brookings Institution Press, 2011).

26. Chuck Lorre and Bill Prady, *The Big Bang Theory*, Chuck Lorre Productions and Warner Bros. (TV series), aired 2007–2019, CBS.

27. Becquerel identified the "phosphorescence" of uranium salts as radiation and, along with Marie and Pierre Curie, won the Nobel Prize in Physics in 1903 for the discovery of radioactivity. Henri Becquerel, "Sur les radiations émises par phosphorescence," *Comptes rendus de 1'Académie des Sciences, Paris* 122 (1896): 420–421.

28. Genovefa Kefalidou and Sarah Sharples, "Encouraging Serendipity in Research: Designing Technologies to Support Connection-Making," *International Journal of Human-Computer Studies* 89 (2016): 1–23.

29. Earth Overshoot Day is calculated by the Geneva Environment Network at: www .genevaenvironmentnetwork.org/resources/updates/earth-overshoot-day/.

30. Gro Harlem Brundtland, "Our Common Future—Call for Action," *Environmental Conservation* 14, no. 4 (1987): 291–294.

31. Charles J. Kahane, *Injury Vulnerability and Effectiveness of Occupant Protection Technologies for Older Occupants and Women*, technical report (2013).

32. Caroline Boudoux, *Fundamentals of Biomedical Optics* (Blurb/Pollux, 2016).

33. Matthew D. Keller et al., *Skin Colour Affects the Accuracy of Medical Oxygen Sensors* 610, no. 7932 (October 19, 2022): 449–451, https://www.nature.com/articles/d41586 -022-03161-1.

34. Shailendra Kumar and Sanghamitra Choudhury, "Gender and Feminist Considerations in Artificial Intelligence from a Developing-World Perspective, with India as a Case Study," *Humanities and Social Sciences Communications* 9, no. 1 (2022): 1–9.

35. Claudia Zaslavsky, "Women as the First Mathematicians," *International Study Groups on Ethnomathematics Newsletter* 7, no. 1 (1992).

36. Chris Welch, "Apple HealthKit Announced: A Hub for All Your iOS Fitness Tracking Needs," *The Verge*, 2014, as reported by Cara Tenenbaum, "Not Intelligent: Encoding Gender Bias," *Minnesota Journal of Law, Science & Technology* 21, no. 2 (2020): 283.

37. Arielle Duhaime-Ross, "Apple Promised an Expansive Health App, So Why Can't I Track Menstruation," *The Verge*, 2014.

CHAPTER 8

1. Aristide van Aartsengel and Selahattin Kurtoglu, "Create Work Breakdown Structure," in *Handbook on Continuous Improvement Transformation* (Berlin Heidelberg: Springer-Verlag, 2013), 137–142.

2. Until you begin mentoring interns, that is.

3. Do not make the mistake of disappearing from the lab during the school year— while good grades are important to obtain scholarships, no one gets a PhD from coursework alone!

4. Snieder and Larner, *The Art of Being a Scientist*.

5. Dwight D. Eisenhower, "Address at the Second Assembly of the World Council of Churches" Evanston, IL, 1954.

6. Meng Zhu, Yang Yang, and Christopher K. Hsee, "The Mere Urgency Effect," *Journal of Consumer Research* 45, no. 3 (February 2018): 673–690.

7. Stephen R. Covey and Sean Covey, *The 7 Habits of Highly Effective People* (Simon & Schuster, 2020).

8. Perhaps unsurprisingly, North America uses different paper sizes than the rest of the world. The American equivalent to A4 is called letter, while the legal format is three *inches* longer, allowing for intimidating to-do lists.

9. Christian W. Dawson, *Projects in Computing and Information Systems: A Student's Guide* (Pearson Education, 2005).

10. Paul Bocuse was a prolific French chef, famous for his frog leg recipes.

11. Mihaly Csikszentmihalyi, *Creativity: Flow and the psychology of discovery and invention* (New York: Harper Perennial, 1997).

12. Jake Knapp and John Zeratsky, *Make Time: How to Focus on What Matters Every Day* (Random House, 2018).

13. Gloria Mark, Daniela Gudith, and Ulrich Klocke, "The Cost of Interrupted Work: More Speed and Stress," in *Proceedings of the SIGCHI conference on Human Factors in Computing Systems* (2008), 107–110.

CHAPTER 9

1. Yves Langhame, "CAP7015E Leading a Research Project," class notes.

2. Bernard-André Genest and Tho Hau Nguyen, *Principes et techniques de la gestion de projets* (Éditions Sigma Delta, 2002).

3. Gregory S. Patience et al., "Intellectual Contributions Meriting Authorship: Survey Results From The Top Cited Authors Across All Science Categories," *PLoS One* 14, no. 1 (2019): e0198117.

4. ICQ was an early instant messaging platform playing on the words *I seek you* with its acronym. I not only reveal my age with this anecdote but also how slow I can be at picking up puns.

5. Fund and Institute, *Making the Right Moves.*

6. Suné Von Solms, Hannelie Nel, and Johan Meyer, "Gender Dynamics: A Case Study of Role Allocation in Engineering Education," *IEEE Access* 6 (2017): 270–279.

7. N. G. Holmes et al., "Evaluating the Role of Student Preference in Physics Lab Group Equity," *Physical Review Physics Education Research* 18, no. 1 (2022): 010106.

8. Pam Roberts and Mary Ayre, "Did She Jump or Was She Pushed? A Study of Women's Retention in the Engineering Workforce," *International Journal of Engineering Education* 18, no. 4 (2002): 415–421.

9. Science digest websites include phys.org, C&EN, or *Nature News.*

10. ISO and ISO, *ISO GUIDE 73: 2009 Risk Management-Vocabulary*, 2009.

11. Paul R. Amyotte and Douglas J. McCutcheon, "Risk Management an Area of Knowledge for All Engineers," *Hg. v. The Research Committee of the Canadian Council of Professional Engineers*, 2006.

12. Prof. Thomas Brzutowski is a past president of NSERC. The Swiss cheese model explains types of research projects Thomas Brzustowski, "IQC Short Workshop on Quantum Entrepreneurship," *Lecture* 3 (2012).

13. John L. Hess and A. M. O. Smith, "Calculation of Potential Flow about Arbitrary Bodies," *Progress in Aerospace Sciences* 8 (1967): 1–138.

14. Marie Kondo, *Spark Joy: An Illustrated Master Class on the Art of Organizing and Tidying Up* (Ten Speed Press, 2016).

15. Paul Bloom, *Just Babies: The Origins of Good and Evil* (Broadway Books, 2013).

16. Minette Drumwright, Robert Prentice, and Cara Biasucci, "Behavioral Ethics and Teaching Ethical Decision Making," *Decision Sciences Journal of Innovative Education* 13, no. 3 (2015): 431–458.

17. *Politique d'éthique et d'intégrité scientifique*, technical report (Fonds québécois de la recherche sur la nature et les technologies, 2010).

18. United States National Commission for the Protection of Human Subjects of Biomedical and Behavioral Research, *The Belmont Report: Ethical Principles and Guidelines for the Protection of Human Subjects of Research*, vol. 2 (The Commission, 1978).

19. "Plagiarize," Merriam-Webster, 2022, https://www.merriam-webster.com.

20. McGill University has published an online quiz to review proper citation techniques. You may take it at https://www.mcgill.ca/students/srr/honest/students/test.

CHAPTER 10

1. Elizabeth Gilbert, *Creative Living Beyond Fear* (New York: Penguin, 2016).

2. Schmich, "Advice, Like Youth, Probably Just Wasted on the Young."

3. Barbara E. Lovitts, "How to Grade a Dissertation," *Academe* 91, no. 6 (2005): 18–23.

4. Louis De Broglie, "Recherches sur la théorie des quanta" (PhD diss., Migration-université en cours d'affectation, 1924).

5. Louis De Broglie, "Waves and Quanta," *Nature* 112, no. 2815 (1923): 540.

6. Staff, "PhD Thesis Structure and Content," University College London, accessed October 16, 2023, http://www0.cs.ucl.ac.uk/staff/C.Clack/phd.html.

7. Luiz Otavio Barros, *The Only Academic Phrasebook You'll Ever Need: 600 Examples of Academic Language* (Bolton, ON: Createspace Independent Publishing Platform, 2016).

8. In Finland, successful candidates get a sword and a hat in addition to their diploma. https://www.discoverphds.com/blog/finlands-phd-sword-and-hat-tradition.

9. However, in the Netherlands, the thesis defense is highly ceremonial and two close friends, called *paranymphs*, stand by your sides, akin to witnesses at a wedding. https://www.gildeprint.nl/en/2020/06/09/role-paranymph-promotion/.

10. *How to Survive Your Viva* by Rowena Murray (ISBN 0-335-21284-0) or *The Doctoral Examination Process: A Handbook for Students, Examiners, and Supervisors* by Penny Tinkler and Carolyn Jackson (ISBN 978-0-335-21305-4).

11. Jorge Cham, How Many Ph.D.'s Does it Take to Get a Powerpoint Presentation to Work ?, 2012.

12. Joannes Barend Klitsie, Rebecca Anne Price, and Christine Stefanie Heleen De Lille, "Overcoming the Valley of Death: A Design Innovation Perspective," *Design Management Journal* 14, no. 1 (2019): 28–41.

13. Laurier L. Schramm, *Technological Innovation: An Introduction* (De Gruyter, 2017).

14. World Intellectual Property Organization, *Understanding Industrial Property*, 2016.

CHAPTER 11

1. Peter J Feibelman, *A PhD is Not Enough!: A Guide to Survival in Science* (Basic Books, 2011).

2. Stephen B. Heard, *The Scientist's Guide to Writing: How to Write More Easily and Effectively throughout Your Scientific Career* (Princeton, NJ: Princeton University Press, 2022).

3. Julia Eckhoff, *Write an Abstract. Springer Nature/Chemistry*, accessed January 18, 2019, https://chemistrycommunity.nature.com/posts/43071-how-to-write-an-abstract.

4. Rajesh S. Pillai et al., "Multiplexed Two-Photon Microscopy of Dynamic Biological Samples with Shaped Broadband Pulses," *Optics Express* 17, no. 15 (2009): 12741–12752.

5. James Hartley and Guillaume Cabanac, "Thirteen Ways to Write an Abstract," *Publications* 5, no. 2 (2017): 11.

6. Chelsea Lee, "An Abbreviations FAQ," October 2015, https://blog.apastyle.org/apastyle/abbreviations/#Q10.

7. A close second is that of a telecentric telescope. Brownie points if you know what this is.

8. Scitable by Nature Education, "Effective Writing," accessed October 16, 2023, https://www.nature.com/scitable/topicpage/effective-writing-13815989/.

9. Boudoux et al., "RapidWavelength-Swept Spectrally Encoded Confocal Microscopy."

10. Along with her doctoral advisor Gérard Mourou, and shared with Arthur Ashkin.

11. Estelle M. Phillips and Derek S. Pugh, "How to Get a Ph.D." (PhD diss., Maidenhead: Open University Press, 2007).

12. Barros, *The Only Academic Phrasebook You'll Ever Need.*

CHAPTER 12

1. David Wade Chambers's research was published in 1983 in David Wade Chambers, "Stereotypic Images of the Scientist: The Draw-a-Scientist Test," *Science Education* 67, no. 2 (1983): 255–265.

2. Kelly R. Stevens et al., "Fund Black Scientists," *Cell* 184, no. 3 (2021): 561–565.

3. Diego F. M. Oliveira et al., "Comparison of National Institutes of Health Grant Amounts to First-Time Male and Female Principal Investigators," *JAMA* 321, no. 9 (2019): 898–900.

4. Gita Ghiasi, Vincent Larivière, and Cassidy Sugimoto, "Gender Differences in Synchronous and Diachronous Self-Citations," in *21st International Conference on Science and Technology Indicators-STI 2016. Book of Proceedings* (2016); Michael H. K. Bendels et al., "Gender Disparities in High-Quality Research Revealed by Nature Index Journals," *PloS One* 13, no. 1 (2018): e0189136; Yiqin Alicia Shen et al., "Persistent Underrepresentation of Women's Science in High Profile Journals," *BioRxiv*, 2018, 275362.

5. Erin A. Cech and T. J. Waidzunas, "Systemic Inequalities for LGBTQ Professionals in STEM," *Science Advances* 7, no. 3 (2021): eabe0933.

6. Trix and Psenka, "Exploring the Color of Glass."

7. As compiled in the meta-analysis: David I. Miller et al., "The Development of Children's Gender-Science Stereotypes: A Meta-Analysis of 5 Decades of US Draw-a-Scientist Studies," *Child Development* 89, no. 6 (2018): 1943–1955 and reported by Ed Yong, "What we Learn From 50 Years of Kids Drawing Scientists," *The Atlantic* 20 (2018), and K. Langin, "What Does a Scientist Look Like? Children are Drawing Women More than Ever Before," *Science* 20 (2018).

8. Inspired from PolyMtl's IDEA policy. https://www.polymtl.ca/edi/politique.

9. Daniel Kahneman et al., *Judgment Under Uncertainty: Heuristics and Biases* (Cambridge: Cambridge University Press, 1982).

10. Pauline Rose Clance, *The Impostor Phenomenon: Overcoming the Fear that Haunts Your Success* (Peachtree Publishers, 1985).

11. Dena M. Bravata et al., "Prevalence, Predictors, and Treatment of Impostor Syndrome: A Systematic Review," *Journal of General Internal Medicine* 35, no. 4 (2020): 1252–1275.

12. Aishwarya Joshi and Haley Mangette, "Unmasking of Impostor Syndrome," *Journal of Research, Assessment, and Practice in Higher Education* 3, no. 1 (2018): 3.

13. Adam Grant, *Think Again: The Power of Knowing What You Don't Know* (Penguin, 2021).

14. Teresa M. Evans et al., "Evidence for a Mental Health Crisis in Graduate Education," *Nature Biotechnology* 36, no. 3 (2018): 282–284.

15. Susan T. Charles, Melissa M. Karnaze, and Frances M. Leslie, "Positive Factors Related to Graduate Student Mental Health," *Journal of American College Health* 70, no. 6 (August–September 2021): 1–9.

16. Jenny K. Hyun et al., "Graduate Student Mental Health: Needs Assessment and Utilization of Counseling Services," *Journal of College Student Development* 47, no. 3 (2006): 247–266; Katia Levecque et al., "Work Organization and Mental Health Problems in PhD Students," *Research Policy* 46, no. 4 (2017): 868–879; Evans et al., "Evidence for a Mental Health Crisis in Graduate Education."

17. Disclaimer adapted from Bork and Mondisa, "Engineering Graduate Students' Mental Health: A Scoping Literature Review."

18. John R. Schermerhorn Jr. et al., *Organizational Behavior* (John Wiley & Sons, 2011).

19. National Research Council, et al., *Committee on Recognition and Alleviation of Distress in Laboratory Animals. Recognition and Alleviation of Distress in Laboratory Animals*, 2008 https://www.ncbi.nlm.nih.gov/books/NBK4027/.

20. Evans et al., "Evidence for a Mental Health Crisis in Graduate Education."

21. Schermerhorn Jr. et al., *Organizational Behavior*.

22. Stevan E. Hobfoll et al., "Conservation of Social Resources: Social Support Resource Theory," *Journal of Social and Personal Relationships* 7, no. 4 (1990): 465–478; and Dan S. Chiaburu, Natalia M. Lorinkova, and Linn Van Dyne, "Employees' Social Context and Change-Oriented Citizenship: A Meta-Analysis of Leader, Coworker, and Organizational Influences," *Group & Organization Management* 38, no. 3 (2013): 291–333.

23. Charles, Karnaze, and Leslie, "Positive Factors Related to Graduate Student Mental Health."

24. Richard Layard and David M. Clark, *Thrive: How Better Mental Health Care Transforms Lives and Saves Money* (Princeton, NJ: Princeton University Press, 2015).

25. Paul Barreira, Matthew Basilico, and Valentin Bolotnyy, "Graduate Student Mental Health: Lessons from American Economics Department," working paper, Harvard University, November 4, 2029, https://scholar.harvard.edu/sites/scholar.harvard.edu/files/bolotnyy/files/bbb_mentalhealth_paper.pdf.

26. Chun-Mei Zhao, Chris M. Golde, and Alexander C. McCormick, "More Than a Signature: How Advisor Choice and Advisor Behaviour Affect Doctoral Student Satisfaction," *Journal of Further and Higher Education* 31, no. 3 (2007): 263–281.

27. Sarah Ketchen Lipson and Daniel Eisenberg, "Mental Health and Academic Attitudes and Expectations in University Populations: Results From the Healthy Minds Study," *Journal of Mental Health* 27, no. 3 (2018): 205–213.

28. Engineering National Academies of Sciences and Medicine, *Mental Health, Substance Use, and Wellbeing in Higher Education: Supporting the Whole Student*, ed. Alan I. Leshner and Layne A. Scherer (Washington, DC: The National Academies Press, 2021), https://doi.org/10.17226/26015.

29. Meri Davlasheridze, Stephan J. Goetz, and Yicheol Han, "The Effect of Mental Health on US County Economic Growth," *Review of Regional Studies* 48, no. 2 (2018): 155–171.

30. Charles, Karnaze, and Leslie, "Positive factors related to graduate student mental health."

31. Barreira, Basilico, and Bolotnyy, "Graduate Student Mental Health."

32. Joanna Hughes, "Managing Your Mental Health as a PhD Student," PhDStudies, July 24, 2019, https://www.phdstudies.com/article/managing-your-mental-health-as -a-phd-student.

33. Sarah Ketchen Lipson et al., "Major Differences: Variations in Undergraduate and Graduate Student Mental Health and Treatment Utilization Across Academic Disciplines," *Journal of College Student Psychotherapy* 30, no. 1 (2016): 23–41.

34. Ståle Einarsen, Merethe Schanke Aasland, and Anders Skogstad, Destructive Leadership Behaviour: A Definition and Conceptual Model, *Leadership Quarterly* 18, no. 3 (2007): 207–216.

35. Christian N. Thoroughgood et al., "Bad to the Bone: Empirically Defining and Measuring Destructive Leader Behavior," *Journal of Leadership & Organizational Studies* 19, no. 2 (2012): 230–255, https://doi.org/10.1177/1548051811436327, eprint: https://doi.org/10.1177/1548051811436327, https://doi.org/10.1177/154805 1811436327.

36. Cornell University Office of the Dean of Faculty, "What Needs to Be Said about Power Differentials?" accessed October 16, 2023, https://theuniversityfaculty.cornell .edu/dean/report-archive/q3-what-about-power-differentials/power-differntials/.

37. Title IX is a US civil rights law enacted as part of the Education Amendments of 1972 that prohibits sex-based discrimination in any education institution that receives federal funding.

38. Marta Burczycka, "Students' Experiences of Unwanted Sexualized Behaviours and Sexual Assault at Postsecondary Schools in the Canadian Provinces," Statistics Canada—Canadian Centre for Justice and Community Safety Statistic, September 2020.

39. Valentina Beghini et al., "Violence and Harassment in the World of Work: A Guide on Convention No. 190 and Recommendation No. 206," Geneva: International Labour Office, 2021, https://www.ilo.org/global/topics/violence-harassment /resources/WCMS_814507/lang–en/index.htm.

40. "Harassment," Merriam-Webster, 2022, https://www.merriam-webster.com.

41. Lynne M. Andersson and Christine M. Pearson., "Tit for Tat? The Spiraling Effect of Incivility in the Workplace," *Academy of Management Review* 24, no. 3 (1999): 452–471.

42. McGill University—Ombudsperson, accessed 16 October 2023, https://www.mc gill.ca/ombudsperson/ and https://accuo.ca/about/a-word-about-the-association/.

43. Grant, *Think Again*, 210.

44. Amy C. Edmondson, *The Fearless Organization: Creating Psychological Safety in the Workplace for Learning, Innovation, and Growth* (John Wiley & Sons, 2018).

45. Grant, *Think Again*.

46. Mentioning the rate here would definitely feel like bragging. Suffice it to say that the daily rate is on the order of magnitude of one single venti nonfat matcha latte, double foam.

CHAPTER 13

1. Grant, *Think Again* 245.

BIBLIOGRAPHY

Aartsengel, Aristide van, and Selahattin Kurtoglu. "Create Work Breakdown Structure." In *Handbook on Continuous Improvement Transformation*, 137–142. Berlin Heidelberg: Springer-Verlag, 2013.

Abraham, Philip, et al. "Duplicate and Salami Publications." *Journal of Postgraduate Medicine* 46, no. 2 (2000): 67.

Adams, Douglas. *The Hitchhiker's Guide to the Galaxy*. UK: Pan Books, 1979.

American Society for Cell Biology et al. "San Francisco Declaration on Research Assessment (DORA)," 2012.

Amyotte, Paul R., and Douglas J. McCutcheon. "Risk Management an Area of Knowledge for All Engineers." *Hg. v. The Research Committee of the Canadian Council of Professional Engineers*, 2006, discussion paper prepared for the Research Committee of the Canadian Council of Professional Engineers, October 2006. https://engineerscanada.ca/sites/default/files/risk_management_paper_eng.pdf.

Andersson, Lynne M., and Christine M. Pearson. "Tit for Tat? The Spiraling Effect of Incivility in the Workplace." *Academy of Management Review* 24, no. 3 (1999): 452–471.

Association du Corps Intermédiaire de l'EPFL. *Practical Guide for PhD Candidates at EPFL*, December 2011. http://acide.epfl.ch/wp-content/uploads/2014/12/PhD_Guide_2011.pdf.

Atkinson, Roger. "Project Management: Cost, Time and Quality, Two Best Guesses and a Phenomenon, Its Time to Accept Other Success Criteria." *International Journal of Project Management* 17, no. 6 (1999): 337–342.

Barker, Kathy *At the Helm: A Laboratory Navigator*. Cold Spring Harbor, NY: Cold Spring Harbor Laboratory Press, 2002.

Barreira, Paul, Matthew Basilico, and Valentin Bolotnyy. "Graduate Student Mental Health: Lessons from American Economics Department." working paper, Harvard University, November 4, 2029. https://scholar.harvard.edu/sites/scholar.harvard.edu /files/bolotnyy/files/bbb_mentalhealth_paper.pdf.

Barros, Luiz Otavio. *The Only Academic Phrasebook You'll Ever Need: 600 Examples of Academic Language*. Bolton, ON: Createspace Independent Publishing Platform, 2016.

Bast, Robert C. "Status of Tumor Markers in Ovarian Cancer Screening." *Supplement. Journal of Clinical Oncology : Official Journal of the American Society of Clinical Oncology* 21, no. 10 (May 2003): 200s–205s.

Morissette, Alanis. *Jagged Little Pill*, 1996.

Becquerel, Henri. "Sur les radiations émises par phosphorescence." *Comptes rendus de l'Académie des Sciences, Paris* 122 (1896): 420–421.

Beghini, Valentina. "Violence and Harassment in the World of Work: A Guide on Convention No. 190 and Recommendation No. 206", Geneva: International Labour Office. 2021. https://www.ilo.org/global/topics/violence-harassment/resources/WC MS_814507/lang--en/index.htm.

Bendels, Michael H. K., Ruth Müller, Doerthe Brueggmann, and David A. Groneberg. "Gender Disparities in High-Quality Research Revealed by Nature Index Journals." *PloS One* 13, no. 1 (2018): e0189136.

Bentley, Jon. *Programming Pearls*. New York: Addison-Wesley Professional, 2016.

Berdanier, Catherine G. P., Carey Whitehair, Adam Kirn, and Derrick Satterfield. "Analysis of Social Media Forums to Elicit Narratives of Graduate Engineering Student Attrition." *Journal of Engineering Education* 109, no. 1 (2020): 125–147.

Björk, Bo-Christer Sari Kanto-Karvonen, and J. Tuomas Harviainen. "How Frequently Are Articles in Predatory Open Access Journals Cited." *Publications* 8, no. 2 (2020): 17.

Bloom, Paul. *Just Babies: The Origins of Good and Evil*. New York: Broadway Books, 2013.

Booth, Andrew, Anthea Sutton, and Diana Papaioannou. *Systematic Approaches to a Successful Literature Review*, 2nd ed. (London: Sage, 2016).

Bork, Sarah Jane, and Joi-Lynn Mondisa. "Engineering Graduate Students' Mental Health: A Scoping Literature Review." *Journal of Engineering Education* 111, no. 3 (2022): 665–702.

Boudoux, Caroline. *Fundamentals of Biomedical Optics*. Blurb/Pollux, 2016.

Boudoux, Caroline. "Wavelength Swept Spectrally Encoded Confocal Microscopy for Biological and Clinical Applications." PhD diss., MIT, 2007.

Boudoux, Caroline, Shelby C. Leuin, Wang Yuhl Oh, Melissa J. Suter, Adrien E. Desjardins, Benjamin J. Vakoc, Brett E. Bouma, Christopher J. Hartnick, and Guillermo J. Tearney. "Optical Microscopy of the Pediatric Vocal Fold." *Archives of Otolaryngology–Head & Neck Surgery* 135, no. 1 (2009): 53–64.

Boudoux, Caroline, Dvir Yelin, Jason T. Motz, Brett E. Bouma, and Guillermo J. Tearney. "Spectral Encoding: A Novel Platform for Endoscopy and Microscopy." In *Laser Science*, JWC5. Optica Publishing Group, 2006. https://opg.optica.org/conference.cfm?meetingid=69&yr=2006#JWC.

Boudoux, Caroline, Seok-Hyun Yun, Wang-Yuhl Oh, W. Matt White, Nicusor V. Iftimia, Milen Shishkov, Brett E. Bouma, and Guillermo J. Tearney. "Rapid Wavelength-Swept Spectrally Encoded Confocal Microscopy." *Optics Express* 13, no. 20 (2005): 8214–8221.

Bravata, Dena M., Sharon A. Watts, Autumn L. Keefer, Divya K. Madhusudhan, Katie T. Taylor, Dani M. Clark, Ross S. Nelson, Kevin O. Cokley, and Heather K. Hagg. "Prevalence, Predictors, and Treatment of Impostor Syndrome: A Systematic Review." *Journal of General Internal Medicine* 35, no. 4 (2020): 1252–1275.

Brent, Alan Colin. "Transdisciplinary Approaches to Engineering R&D: Importance of Understanding Values and Culture." In *Handbook of Sustainable Engineering*, edited by Joanne Kauffman and Lee Kun-Mo. Netherlands: Springer, 2013.

Brundtland, Gro Harlem. "Our Common Future—Call for Action." *Environmental Conservation* 14, no. 4 (1987): 291–294.

Brzustowski, Thomas. "IQC Short Workshop on Quantum Entrepreneurship." *Lecture* 3 (2012).

Burczycka, Marta. Students' Experiences of Unwanted Sexualized Behaviours and Sexual Assault at Postsecondary Schools in the Canadian Provinces. Statistics Canada—Canadian Centre for Justice and Community Safety Statistic, September 2020.

Bush, Vanevar. *Science, the Endless Frontier*. Washington: National Science Foundation–EUA, 1945.

Cech, Erin A., and T. J. Waidzunas. "Systemic Inequalities for LGBTQ Professionals in STEM." *Science Advances* 7, no. 3 (2021): eabe0933.

Cham, Jorge. *How Many Ph.D.'s Does it Take to Get a Powerpoint Presentation to Work?*, 2012.

Cham, Jorge. *The Thesis Committee*. PhD Comics, November 16, 2012. https://phdcomics.com/comics/archive_print.php?comicid=1537.

Chambers, David Wade. "Stereotypic Images of the Scientist: The Draw-a-Scientist Test." *Science Education* 67, no. 2 (1983): 255–265.

Charles, Susan T, Melissa M, Karnaze, and Frances M. Leslie. "Positive Factors Related to Graduate Student Mental Health." *Journal of American College Health* 70, no. 6 (2021): 1858–1866. https://doi.org/10.1080/07448481.2020.1841207.

Chiaburu, Dan S., Natalia M. Lorinkova, and Linn Van Dyne. "Employees' Social Context and Change-Oriented Citizenship: A Meta-Analysis of Leader, Coworker, and Organizational Influences." *Group & Organization Management* 38, no. 3 (2013): 291–333.

Choi, Bernard C. K., and Anita W. P. Pak. "Multidisciplinarity, Interdisciplinarity, and Transdisciplinarity in Health Research, Services, Education and Policy: 2. Promotors, Barriers, and Strategies of Enhancement." *Clinical and Investigative Medicine* 30, no. 6 (2007): E224–E232.

Clance, Pauline Rose. *The Impostor Phenomenon: Overcoming the Fear that Haunts Your Success*. Peachtree Publishers, 1985.

Cleland, David I., and William R. King. *Project Management Handbook*. New York: Van, 1988.

Conroy, Gemma. "What's Wrong with the H-index, According to Its Inventor." *Nature Index* 24, 2020.

Cornér, Solveig, Erika Löfström, and Kirsi Pyhältö. "The Relationship Between Doctoral Students' Perceptions of Supervision and Burnout." *International Journal of Doctoral Studies* 12 (2017): 91–106. https://doi.org/10.28945/3754.

Covey, Stephen R., and Sean Covey. *The 7 Habits of Highly Effective People*. Simon & Schuster, 2020.

Crane, David Crane, and Marta Kauffman. *Friends*, 1994.

Cray, Heather. "How to Make an Original Contribution to Knowledge." *University Affairs/Affaires Universitaires*, August 2014.

Cretton, Destin Daniel. *Shang-Chi and the Legend of the Ten Rings*. Marvel Studios, 2021.

Crouch, Catherine, Adam P. Fagen, J. Paul Callan, and Eric Mazur. "Classroom Demonstrations: Learning Tools or Entertainment?" *American Journal of Physics* 72, no. 6 (2004): 835–838.

Crouch, Catherine H., and Eric Mazur. "Peer Instruction: Ten Years of Experience and Results." *American Journal of Physics* 69, no. 9 (2001): 970–977.

Cruz, Juan M., Mayra S. Artiles, Holly M. Matusovich, Gwen Lee-Thomas, and Stephanie G. Adams. "Revising the Dissertation Institute: Contextual Factors Relevant to Transferability." In *2019 ASEE Annual Conference & Exposition*, 2019.

Csikszentmihalyi, Mihaly. *Flow and the Psychology of Discovery and Invention*. Harper Perennial, 1997.

Daempfle, Peter A. "An analysis of the High Attrition Rates Among First Year College Science, Math, and Engineering Majors." *Journal of College Student Retention: Research, Theory & Practice* 5, no. 1 (2003): 37–52.

Dansereau, Jean, et al. *Competencies, Competency Elements, and Resources to Mobilize for Graduate Studies, Professional Master's, Research-Based Master's, and Doctorate.* (Montréal: Polytechnique Montréal, 2014). https://www.polymtl.ca/renseignements -generaux/en/official-documents.

Davlasheridze, Meri, Stephan J. Goetz, and Yicheol Han. "The Effect of Mental Health on US County Economic Growth." *Review of Regional Studies* 48, no. 2 (2018): 155–171.

Dawson, Christian W. *Projects in Computing and Information Systems: A Student's Guide.* Pearson Education, 2005.

Day, Robert A., and Barbara Gastel. *How to Write and Publish a Scientific Paper.* 8th ed. Cambridge: Cambridge University Press, 2016.

De Broglie, Louis. "Recherches sur la théorie des quanta." PhD diss., Migration-université en cours d'affectation, 1924.

De Broglie, Louis. "Waves and Quanta." *Nature* 112, no. 2815 (1923): 540.

Denk, Winfried James H. Strickler, and W. Watt Webb. "Two-Photon Laser Scanning Fluorescence Microscopy." *Science* 248, no. 4951 (1990): 73–76.

Desjardins, Patrick. "CAP7003E Doctoral Strategies in Engineering." Class notes.

Diamond, Jared. *Guns, Germs, and Steel.* W. W. Norton, 1997.

Dirac, Paul Adrien Maurice. *Quantum Mechanics.* PhD Thesis. Cambridge: Cambridge University Press, 1926.

Donnelly, Tim. "Brilliant Inventions Made by Mistake." *Inc.* 24 (2012). https://www .inc.com/tim-donnelly/brilliant-failures/9-inventions-made-by-mistake.html.

Doran George T. et al. "There's a SMART Way to Write Management's Goals and Objectives." *Management Review* 70, no. 11 (1981): 35–36.

Drumwright, Minette, Robert Prentice, and Cara Biasucci. "Behavioral Ethics and Teaching Ethical Decision Making." *Decision Sciences Journal of Innovative Education* 13, no. 3 (2015): 431–458.

Dudley, John M. "Defending Basic Research." *Nature Photonics* 7, no. 5 (2013): 338–339.

Duhaime-Ross, Arielle. "Apple Promised an Expansive Health App, So Why Can't I Track Menstruation." *The Verge*, 2014.

Durrani, Matin. "Ig Nobels Prove to be More than a Joke." *Physics World* 16, no. 4 (2003): 12.

Eckhoff, Julia. "How to write an Abstract." Springer Nature/Chemistry, January 18, 2019. https://chemistrycommunity.nature.com/posts/43071-how-to-write-an-abstract.

Eco, Umberto. *How to Write a Thesis.* MIT Press, 2015.

Eddington, A. S. "The Internal Constitution of the Stars." *Nature* 106, no. 2653 (1920): 14–20. https://doi.org/10.1038/106014a0. https://doi.org/10.1038/106014a0.

Editorial. "Time to Remodel the Journal Impact Factor." *Nature* 535, no. 466 (2016).

Edmondson, Amy C. *The Fearless Organization: Creating Psychological Safety in the Workplace for Learning, Innovation, and Growth.* John Wiley & Sons, 2018.

Eisenhower, Dwight D. "Address at the Second Assembly of the World Council of Churches." 1954.

Erickson, Britt K., Michael G. Conner, and Charles N. Landen Jr. "The Role of the Fallopian Tube in the Origin of Ovarian Cancer." *American Journal of Obstetrics and Gynecology* 209, no. 5 (2013): 409–414.

Evans, Teresa M., Lindsay Bira, Jazmin Beltran Gastelum, L. Todd Weiss, and Nathan L. Vanderford. "Evidence for a Mental Health Crisis in Graduate Education." *Nature Biotechnology* 36, no. 3 (2018): 282–284.

Feibelman, Peter J. *A PhD is Not Enough!: A Guide to Survival in Science.* Basic Books, 2011.

Feynman, Richard Phillips. *"Surely You're Joking, Mr. Feynman!": Adventures of a Curious Character.* W. W. Norton, 1985.

Fiske, Peter. "For Your Information." *Nature* 538, no. 7625 (2016): 417–418.

Fiske, Peter S. *Put Your Science to Work: The Take-Charge Career Guide for Scientists.* John Wiley & Sons, 2013.

Fund, Burroughs Wellcome, and Howard Hughes Medical Institute. *Making the Right Moves. A Practical Guide to Scientific Management for Postdocs and New Faculty.* 2006.

Galloway, Scott. *The Algebra of Happiness: Notes on the Pursuit of Success, Love, and Meaning.* Penguin, 2019.

Geisinger, Brandi N., and D. Raj Raman. "Why They Leave: Understanding Student Attrition from Engineering Majors." *International Journal of Engineering Education* 29, no. 4 (2013): 914–925.

Genest, Bernard-André, and Tho Hau Nguyen. *Principes et techniques de la gestion de projets.* Éditions Sigma Delta, 2002.

Ghiasi, Gita, Vincent Larivière, and Cassidy Sugimoto. "Gender Differences in Synchronous and Diachronous Self-Citations." In *21st International Conference on Science and Technology Indicators-STI 2016. Book of Proceedings.* 2016.

Gilbert, Elizabeth. *Big magic: Creative living beyond fear.* Penguin, 2016.

Giltner, David M. *It's a Game Not a Formula: How to Succeed as a Scientist Working in the Private Sector.* SPIE Press, 2021.

Giltner, David M. *Turning Science into Things People Need: Voices of Scientists Working in Industry.* Wide Media Group, 2017.

Goeppert Mayer, Maria. "Elementary Processes with Two Quantum Jumps." *Annalen der Physik* 9 (1931): 273–294.

Goss, Charles Mayo. "Gray's Anatomy of the Human Body." *Academic Medicine* 35, no. 1 (1960): 90.

Grant, Adam. *Think Again: The Power of Knowing What You Don't Know.* Penguin, 2021.

Guier, William H., and George C. Weiffenbach. "Genesis of Satellite Navigation." *Johns Hopkins APL Technical Digest* 18, no. 2 (1997): 179.

Harari, Yuval Noah. *Sapiens.* Bazarforlag AS, 2016.

Hart, Chris. "Doing a Literature Review: Releasing the Research Imagination," 2018.

Hartley, James, and Guillaume Cabanac. "Thirteen Ways to Write an Abstract." *Publications* 5, no. 2 (2017): 11.

Héder, Mihály. "From NASA to EU: The Evolution of the TRL Scale in Public Sector Innovation." *The Innovation Journal* 22, no. 2 (2017): 1–23.

Hergé. *Les bijoux de la Castafiore.* Casterman, 1963.

Hess, John L., and A. M. O. Smith. "Calculation of Potential Flow about Arbitrary Bodies." *Progress in Aerospace Sciences* 8 (1967): 1–138.

Hicks, Diana, Paul Wouters, Ludo Waltman, Sarah De Rijcke, and Ismael Rafols. "Bibliometrics: The Leiden Manifesto for Research Metrics." *Nature* 520, no. 7548 (2015): 429–431.

Hirsch, Jorge E. "An Index to Quantify an Individual's Scientific Research Output." *Proceedings of the National Academy of Sciences* 102, no. 46 (2005): 16569–16572.

Hirsch, Jorge E. "Does the H Index Have Predictive Power?" *Proceedings of the National Academy of Sciences* 104, no. 49 (2007): 19193–19198.

Hobfoll, Stevan E., John Freedy, Carol Lane, and Pamela Geller. "Conservation of Social Resources: Social Support Resource Theory." *Journal of Social and Personal Relationships* 7, no. 4 (1990): 465–478.

Hockney, David, and Charles M. Falco. "Optical Insights into Renaissance Art." *Optics and Photonics*, no. 7 (July 2000): 52–59.

Holmes, N. G., Grace Heath, Katelynn Hubenig, Sophia Jeon, Z. Yasemin Kalender, Emily Stump, and Eleanor C. Sayre. "Evaluating the Role of Student Preference in

Physics Lab Group Equity." *Physical Review Physics Education Research* 18, no. 1 (2022): 010106.

Hou, Xucheng, Tal Zaks, Robert Langer, and Yizhou Dong. "Lipid Nanoparticles for mRNA Delivery." *Nature Reviews Materials* 6, no. 12 (2021): 1078–1094.

Hyun, Jenny K., Brian C. Quinn, Temina Madon, and Steve Lustig. "Graduate Student Mental Health: Needs Assessment and Utilization of Counseling Services." *Journal of College Student Development* 47, no. 3 (2006): 247–266.

ISO and ISO. *ISO GUIDE 73: 2009 Risk Management-Vocabulary*. 2009.

Jarlskog, Cecilia. "Lord Rutherford of Nelson, His 1908 Nobel Prize in Chemistry, and Why He didn't Get a Second Prize." *Journal of Physics: Conference Series* 136, no. 1 (2008): 012001.

Johnson, Carla C., Margaret J. Mohr-Schroeder, Tamara J. Moore, and Lyn D. English. *Handbook of Research on STEM Education*. London: Routledge, 2020.

Joshi, Aishwarya, and Haley Mangette. "Unmasking of Impostor Syndrome." *Journal of Research, Assessment, and Practice in Higher Education* 3, no. 1 (2018): 3.

Kahane, Charles J. *Injury Vulnerability and Effectiveness of Occupant Protection Technologies for Older Occupants and Women*. Technical Report, 2013.

Kahn, Shulamit, and Donna K. Ginther. "The Impact of Postdoctoral Training on Early Careers in Biomedicine." *Nature Biotechnology* 35, no. 1 (2017): 90–94.

Kahneman, Daniel, Stewart Paul Slovic, Paul Slovic, and Amos Tversky. *Judgment Under Uncertainty: Heuristics and Biases*. Cambridge: Cambridge University Press, 1982.

Kaiser, W., and C. G. B. Garrett. "Two-Photon Excitation in Ca F 2: Eu 2+." *Physical Review Letters* 7, no. 6 (1961): 229.

Kang, Dongkyun, Melissa J. Suter, Caroline Boudoux, Patrick S. Yachimski, Norman S. Nishioka, Mari Mino-Kenudson, Gregory Y. Lauwers, Brett E. Bouma, and Guillermo J. Tearney. "Combined Reflection Confocal Microscopy and Optical Coherence Tomography Imaging of Esophageal Biopsy." *Gastrointestinal Endoscopy* 69, no. 5 (2009): AB368.

Kefalidou, Genovefa, and Sarah Sharples. "Encouraging Serendipity in Research: Designing Technologies to Support Connection-Making." *International Journal of Human-Computer Studies* 89 (2016): 1–23.

Keller, Matthew D., Brandon Harrison-Smith, Chetan Patil, and Mohammed Shahriar Arefin. *Skin Colour Affects the Accuracy of Medical Oxygen Sensors* 610, no. 7932 (October 19, 2022): 449–451. https://www.nature.com/articles/d41586-022-03161-1.

King, Margaret F. *Ph.D. Completion and Attrition: Analysis of Baseline Demographic Data from the Ph.D. Completion Project*. Conroe, TX: Nicholson, 2008.

King, Stephen. *On Writing: A Memoir of the Craft*. Simon & Schuster, 2000.

Klitsie, Joannes Barend, Rebecca Anne Price, and Christine Stefanie Heleen De Lille. "Overcoming the Valley of Death: A Design Innovation Perspective." *Design Management Journal* 14, no. 1 (2019): 28–41.

Knapp, Jake, and John Zeratsky. *Make Time: How to Focus on What Matters Every Day*. Random House, 2018.

Kondo, Marie. *Spark Joy: An Illustrated Master Class on the Art of Organizing and Tidying Up*. Ten Speed Press, 2016.

Kumar, Shailendra, and Sanghamitra Choudhury. "Gender and Feminist Considerations in Artificial Intelligence from a Developing-World Perspective, with India as a Case Study." *Humanities and Social Sciences Communications* 9, no. 1 (2022): 1–9.

Langhame, Yves. "CAP7015E Leading a Research Project." Class notes, 2015.

Langin, Katie. "What Does a Scientist Look Like? Children Are Drawing Women More than Ever Before." *Science*, March 20, 2018. https://www.science.org/content/article/what-does-scientist-look-children-are-drawing-women-more-ever.

Langin, Katie. "It's OK to Quit Your Ph.D." *Science*, June 25, 2019. https://www.science.org/content/article/it-s-ok-quit-your-phd.

Lariviere, Vincent, Veronique Kiermer, Catriona J. MacCallum, Marcia McNutt, Mark Patterson, Bernd Pulverer, Sowmya Swaminathan, Stuart Taylor, and Stephen Curry. "A Simple Proposal for the Publication of Journal Citation Distributions." *BioRxiv*, 2016, 062109.

Lariviere, Vincent, and Cassidy R. Sugimoto. "The Journal Impact Factor: A Brief History, Critique, and Discussion of Adverse Effects." In *Springer Handbook of Science and Technology Indicators*, edited by Wolfgang Glänzel, Henk F. Moed, Ulrich Schmoch, and Mike Thelwall, 3–24. Springer, 2019.

Larivière, Vincent. "PhD Students' Excellence Scholarships and Their Relationship with Research Productivity, Scientific Impact, and Degree Completion." *Canadian Journal of Higher Education* 43, no. 2 (2013): 27–41.

Layard, Richard, and David M. Clark. *Thrive: How Better Mental Health Care Transforms Lives and Saves Money*. Princeton, NJ: Princeton University Press, 2015.

Lee, Chelsea. "An Abbreviations FAQ," October 2015. https://blog.apastyle.org/apastyle/abbreviations/#Q10.

Leone, Sergio. *The Good, the Bad and the Ugly*. Produzioni Europee Associate, 1966.

Levecque, Katia, Frederik Anseel, Alain De Beuckelaer, Johan Van der Heyden, and Lydia Gisle. "Work Organization and Mental Health Problems in PhD Students." *Research Policy* 46, no. 4 (2017): 868–879.

Ling, Charles X., and Qiang Yang. "Crafting Your Research Future: A Guide to Successful Master's and Ph.D. Degrees in Science & Engineering." *Synthesis Lectures on Engineering* 7, no. 3 (2012): 1–168.

Lipson, Sarah Ketchen, and Daniel Eisenberg. "Mental Health and Academic Attitudes and Expectations in University Populations: Results From the Healthy Minds Study." *Journal of Mental Health* 27, no. 3 (2018): 205–213.

Lipson, Sarah Ketchen, Sasha Zhou, Blake Wagner III, Katie Beck, and Daniel Eisenberg. "Major Differences: Variations in Undergraduate and Graduate Student Mental Health and Treatment Utilization Across Academic Disciplines." *Journal of College Student Psychotherapy* 30, no. 1 (2016): 23–41.

Lorre, Chuck, and Bill Prady. *The Big Bang Theory* Chuck Lorre Productions and Warner Bros. (TV series), aired 2007-2019, CBS.

Lovitts, Barbara E. "How to Grade a Dissertation." *Academe* 91, no. 6 (2005): 18–23.

Lovitts, Barbara E. *Making the Implicit Explicit: Creating Performance Expectations for the Dissertation.* Stylus Publishing, 2007.

Merriam-Webster Staff. *Merriam-Webster's Collegiate Dictionary.* Vol. 2. Merriam-Webster, 2004.

Mark, Gloria, Daniela Gudith, and Ulrich Klocke. "The Cost of Interrupted Work: More Speed and Stress." In *Proceedings of the SIGCHI conference on Human Factors in Computing Systems,* 2008: 107–110.

Maslow, Abraham H. "A Dynamic Theory of Human Motivation." *Psychological Review* 50, no. 4 (1958): 370–396. https://doi.org/10.1037/h0054346.

Might, Matt. *The Illustrated Guide to a Ph.D.* November 2020. http://matt.might.net /articles/phd-school-in-pictures.

Miller, David I., Kyle M. Nolla, Alice H. Eagly, and David H. Uttal. "The Development of Children's Gender-Science Stereotypes: A Meta-Analysis of 5 Decades of US Draw-a-Scientist Studies." *Child Development* 89, no. 6 (2018): 1943–1955.

"MIT Can't Read or Harvard Can't Count Joke." *The Los Angeles Times,* October 23, 1966.

Momenimovahed, Zohre, Azita Tiznobaik, Safoura Taheri, and Hamid Salehiniya. "Ovarian Cancer in the World: Epidemiology and Risk Factors." *International Journal of Women's Health* 11 (2019): 287.

National Academies of Sciences, Engineering, and Medicine. *Mental Health, Substance Use, and Wellbeing in Higher Education: Supporting the Whole Student.* Edited by Alan I. Leshner and Layne A. Scherer. Washington, DC: The National Academies Press, 2021. https://doi.org/10.17226/26015.

National Research Council, et al. *Committee on Recognition and Alleviation of Distress in Laboratory Animals. Recognition and Alleviation of Distress in Laboratory Animals.* https://www.ncbi.nlm.nih.gov/books/NBK4027/.

Nicolas, Jean. "Réussir son doctorat". class notes, 2012.

Oliveira, Diego F. M., Yifang Ma, Teresa K. Woodruff, and Brian Uzzi. "Comparison of National Institutes of Health Grant Amounts to First-Time Male and Female Principal Investigators." *JAMA* 321, no. 9 (2019): 898–900.

Opsomer, J., A. Chen, W. Y. Chang, and D. Foley. *US Employment Higher in the Private Sector than in the Education Sector for US-Trained Doctoral Scientists and Engineers: Findings from the 2019 Survey of Doctorate Recipients.* NSF 21-319, 2021.

Organizatio, International Labour. *Violence and harassment at work: A practical guide for employers,* 2022.

Patience, Gregory S., Daria C. Boffito, and Paul Patience. *Communicate Science Papers, Presentations, and Posters Effectively.* Academic Press, 2015.

Patience, Gregory S., Federico Galli, Paul A. Patience, and Daria C. Boffito. "Intellectual Contributions Meriting Authorship: Survey Results From The Top Cited Authors Across All Science Categories." *PLoS One* 14, no. 1 (2019): e0198117.

Phillips, Estelle M., and Derek S. Pugh. "How to Get a Ph.D." PhD diss., Maidenhead: Open University Press, 2007.

Pillai, Rajesh S., Caroline Boudoux, Guillaume Labroille, Nicolas Olivier, Israel Veilleux, Emmanuel Farge, Manuel Joffre, and Emmanuel Beaurepaire. "Multiplexed Two-Photon Microscopy of Dynamic Biological Samples with Shaped Broadband Pulses." *Optics Express* 17, no. 15 (2009): 12741–12752.

Poinsinet de Sivry-Houle, Martin, Simon Bolduc Beaudoin, Simon Brais-Brunet, Mathieu Dehaes, Nicolas Godbout, and Caroline Boudoux. "All-Fiber Few-Mode Optical Coherence Tomography Using a Modally-Specific Photonic Lantern." *Biomedical Optics Express* 12, no. 9 (2021): 5704–5719.

Politique d'éthique et d'intégrité scientifique. Technical report. Fonds québécois de la recherche sur la nature et les technologies, 2010.

Powell, Kendall. "How to Sail Smoothly from Academia to Industry." *Nature* 555, no. 7697 (2018).

Reitz, Joan M. *Online Dictionary for Library and Information Science.* Danbury, CT: Western Connecticut State University, 1996.

Roberts, Pam, and Mary Ayre. "Did She Jump or Was She Pushed? A Study of Women's Retention in the Engineering Workforce." *International Journal of Engineering Education* 18, no. 4 (2002): 415–421.

Rooij, Els van, Marjon Fokkens-Bruinsma, and E. Jansen. "Factors that Influence PhD Candidates' Success: The Importance of PhD Project Characteristics." *Studies in Continuing Education* 43, no. 12 (2019): 1–20.

Sadlak, Jan. *Doctoral Studies and Qualifications in Europe and the United States: Status and Prospects*. UNESCO, 2004.

Schermerhorn Jr., John R., Richard N. Osborn, Mary Uhl-Bien, and James G. Hunt. *Organizational Behavior*. John Wiley & Sons, 2011.

Schillebeeckx, Maximiliaan, Brett Maricque, and Cory Lewis. "The Missing Piece to Changing the University Culture." *Nature Biotechnology* 31, no. 10 (2013): 938–941.

Schmich, Mary. "Advice, Like Youth, Probably Just Wasted on the Young (Wear Sunscreen)." *Chicago Tribune*, 1997.

Schramm, Laurier L. *Technological Innovation: An Introduction*. De Gruyter, 2017.

Sharp, John A., John Peters, and Keith Howard. *The Management Of a Student Research Project*. Gower Publishing, 2012.

Shen, Yiqin Alicia, Jason M. Webster, Yuichi Shoda, and Ione Fine. "Persistent Underrepresentation of Women's Science in High Profile Journals." *BioRxiv*, 2018, 275362.

Smith, James. "Article title." (City) 14, no. 6 (March 2013): 1–8.

Snieder, Roel, and Ken Larner. *The Art of Being a Scientist: A Guide for Graduate Students And Their Mentors*. Cambridge: Cambridge University Press, 2009.

Stefan, Melanie. "A CV of Failures." *Nature* 468, no. 7322 (2010): 467.

Stevens, Kelly R., Kristyn S. Masters, P. I. Imoukhuede, Karmella A. Haynes, Lori A. Setton, Elizabeth Cosgriff-Hernandez, Muyinatu A. Lediju Bell, et al. "Fund Black Scientists." *Cell* 184, no. 3 (2021): 561–565.

Stokes, Donald E. *Pasteur's Quadrant: Basic Science and Technological Innovation*. Brookings Institution Press, 2011.

Strupler, Mathias, Etienne De Montigny, Dominic Morneau, and Caroline Boudoux. "Rapid Spectrally Encoded Fluorescence Imaging Using a Wavelength-Swept Source." *Optics Letters* 35, no. 11 (2010): 1737–1739.

Suber, Peter. *Open Access*. MIT Press, 2012.

Tarver, Evan. *Corporate Culture*. Investopedia, April 25, 2023. https://www.investopedia.com/terms/c/corporate-culture.asp.

Tenenbaum, Cara. "Not Intelligent: Encoding Gender Bias." *Minnesota Journal of Law, Science & Technology* 21, no. 2 (2020): 283.

Thiel, David V. *Research Methods for Engineers*. Cambridge University Press, 2014.

Think.Check.Submit. https://thinkchecksubmit.org/journals/.

Thoroughgood, Christian N., Brian W. Tate, Katina B. Sawyer, and Rick Jacobs. "Bad to the Bone: Empirically Defining and Measuring Destructive Leader Behavior." *Journal of Leadership & Organizational Studies* 19, no. 2 (2012): 230–255. https://doi .org/10.1177/1548051811436327. eprint: https://doi.org/10.1177/15480518114363 27. https://doi.org/10.1177/1548051811436327.

Tolkien, John Ronald Reuel. *The Lord of the Rings*. Allen & Unwin, 1954.

Tortora, G. J., and B. Derrickson. *The Principles of Anatomy and Physiology*. 15th ed. Wiley, 2017.

Trix, Frances, and Carolyn Psenka. "Exploring the Color of Glass: Letters of Recommendation For Female and Male Medical Faculty." *Discourse & Society* 14, no. 2 (2003): 191–220.

Turing, Alan M. "Computing Machinery and Intelligence." *Mind* 58, no. 236 (1950): 433–460.

United States National Commission for the Protection of Human Subjects of Biomedical and Behavioral Research. *The Belmont Report: Ethical Principles and Guidelines for the Protection of Human Subjects of Research*. Vol. 2. The Commission, 1978.

Van Aken, David C., and William F. Hosford. *Reporting Results: A Practical Guide for Engineers and Scientists*. Cambridge University Press, 2008.

Van Campenhoudt, Luc, Jacques Marquet, and Raymond Quivy. *Manuel de recherche en sciences sociales-5e éd*. Dunod, 2017.

Van Noorden, Richard, and Dalmeet Singh Chawla. "Hundreds of Extreme Self-Citing Scientists Revealed in New Database." *Nature* 572, no. 7771 (2019): 578–580.

Von Der Weid J. P., Rogiero Passy, G. Mussi, and Nicolas Gisin. "On the Characterization of Optical Fiber Network Components with Optical Frequency Domain Reflectometry." *Journal of Lightwave Technology* 15, no. 7 (1997): 1131–1141.

Von Solms, Suné, Hannelie Nel, and Johan Meyer. "Gender Dynamics: A Case Study of Role Allocation in Engineering Education." *IEEE Access* 6 (2017): 270–279.

Wachowskis, The. *The Matrix (movie)*. Warner Bros., 1999.

Welch, Chris. "Apple HealthKit Announced: A Hub for All Your iOS Fitness Tracking Needs." *The Verge*, 2014.

Wigner, Eugene P. "Maria Goeppert Mayer." *Physics Today* 25, no. 5 (May 1972): 77–79.

World Intellectual Property Organization. *Understanding Industrial Property*. 2016.

Yelin, Ronit, Dvir Yelin, Wang-Yuhl Oh, Seok H. Yun, Caroline Boudoux, Benjamin J. Vakoc, Brett E. Bouma, and Guillermo J. Tearney. "Multimodality Optical Imaging of Embryonic Heart Microstructure." *Journal of Biomedical Optics* 12, no. 6 (2007): 064021–064021.

Yemm, Graham. *Financial Times – Essential Guides to Leading Your Team: How To Set Goals, Measure Performance And Reward Talent.* Pearson UK, 2012.

Yong, Ed. "What we Learn From 50 Years of Kids Drawing Scientists." *The Atlantic* 20 (March 20, 2018). https://www.theatlantic.com/science/archive/2018/03/what-we-learn-from-50-years-of-asking-children-to-draw-scientists/556025/.

Yun, Seok-Hyun, C. Boudoux, Guillermo J. Tearney, and Brett E. Bouma. "High-Speed Wavelength-Swept Semiconductor Laser with a Polygon-Scanner-Based Wavelength Filter." *Optics Letters* 28, no. 20 (2003): 1981–1983.

Yun, Seok-Hyun, Caroline Boudoux, Guillermo J. Tearney, and Brett E. Bouma. "Extended-Cavity Semiconductor Wavelength-Swept Laser for Biomedical Imaging." *IEEE Photonics Technology Letters* 16, no. 1 (2004): 293–295.

Zack, Devora. *Networking for People Who Hate Networking: A Field Guide for Introverts, the Overwhelmed, and the Underconnected.* Berrett-Koehler Publishers, 2019.

Zaslavsky, Claudia. "Women as the First Mathematicians." *Newsletter of Women in Mathematics Education* 14, no. 1 (Fall 1991): 4).

Zerbe, Ellen, Gabriella M. Sallai, Kanembe Shanachilubwa, and Catherine G. P. Berdanier. "Engineering Graduate Students' Critical Events as Catalysts of Attrition." *Journal of Engineering Education* 111, no. 4 (2022): 868–888.

Zhao, Chun-Mei, Chris M. Golde, and Alexander C. McCormick. "More Than a Signature: How Advisor Choice and Advisor Behaviour Affect Doctoral Student Satisfaction." *Journal of Further and Higher Education* 31, no. 3 (2007): 263–281.

Zhu, Meng, Yang Yang, and Christopher K. Hsee. "The Mere Urgency Effect." *Journal of Consumer Research* 45, no. 3 (February 2018): 673–690.

CONTRIBUTORS

Professor Sofiane Achiche Department of Mechanical Engineering, PolyMtl, Québec, Canada

Élise Anne Basque librarian, PolyMtl, Québec, Canada

Professor Brett E. Bouma Wellman Center for Photomedicine, Harvard Medical School, MA, US

Doctor Normand Brais Polymath, Sanuvox, Ville St-Laurent, Québec, Canada

Wanessa Cardoso de Sousa Université de Montréal, Québec, Canada

Patrick Cigana Office of Sustainable Development, PolyMtl, Québec, Canada

Ray Daher Office of intervention and prevention of conflicts and violence, from the French *Bureau d'intervention et de prévention des conflits et de la violence* (BIPCV), PolyMtl, Québec, Canada

Professor Patrick Desjardins Department of Engineering Physics, PolyMtl, Québec, Canada

Guylaine Dubreuil Career Management Services, PolyMtl, Québec, Canada

Amel Elimam BIPCV, PolyMtl, Québec, Canada

Professor Virginie Francoeur Department of Industrial Engineering and Applied Mathematics, PolyMtl, Québec, Canada

Professor Jérôme Genest Department of Physics and Engineering Physics, *Université Laval*, Québec, Canada

María-Gracia Girardi   IDEA Advisor, PolyMtl, Québec, Canada

Doctor Felipe Gohring de Magalhães   Department of Computer Engineering, PolyMtl, Québec, Canada

Professor Blanca Himes   Perelman School of Medicine, University of Pennsylvania, PA, US

Professor Éric Laurendeau   Department of Mechanical Engineering, PolyMtl, Québec, Canada

Professor Martin Leahy   Department of Physics, *Ollscoil na Gaillimhe*, Galway, Ireland

Doctor Jiawen Li   Faculty of Sciences, Engineering and Technology, The University of Adelaide, South Australia, Australia

Doctor Éric Proietti   Office of Research, PolyMtl, Québec, Canada

Prof. Derryck T. Reid   Professor of Physics, Heriot-Watt University, Edinburgh, UK

Doctor Élise St-Jacques   Registrar's Office, PolyMtl, Québec, Canada

Professor Jason Robert Tavares   Department of Chemical Engineering, PolyMtl, Québec, Canada

Andrei Uglar   Entrepreneur, Québec, Canada

Doctor Minea Valle Fajer   Office of Graduate Studies, PolyMtl, Québec, Canada

Professor Lucien Weiss   Department of Engineering Physics, PolyMtl, Québec, Canada

AUTHOR INDEX

TOPICS INDEX